G. McDaniel
11-91

CHEVROLET 1955

Creating the Original

by Michael Lamm

Lamm-Morada Inc.
P.O. Box 7607
Stockton, CA 95207

For JoAnne

Distributed by

Motorbooks International
Publishers & Wholesalers Inc
Osceola, Wisconsin 54020, USA ®

ISBN 0-932128-06-8

Library of Congress Catalog Card Number 91-90449

CONTENTS

The 1955 Chevrolet's Development:

AN OVERVIEW

This book tells the story of the making—the creation—of the 1955 Chevrolet. It's told in interview form by the people who actually designed, engineered, and built the car. I had the good fortune, in the mid-1970s and again in 1991, to talk to most of the key figures in the '55 Chevy's development.

It's extremely rare that an entire car gets done from scratch, but that's pretty much what happened in the case of the 1955 Chevrolet. Of the car's 4500 components, all but 675 were newly created for 1955. The three major carryovers were the 6-cylinder engine, the manual transmission, and the basic Powerglide automatic. But the frame, suspension, rear axle, body, accessories, and especially the V8 engine...those and much more were totally new.

Four things stand out in the 1955 Chevrolet's development. First, it all filtered through one man: Edward Nicholas Cole. Without Ed Cole's organizational skill, his willingness to try the untried, and his personality and ability to motivate, the 1955 Chevrolet wouldn't have been nearly the car it became. Cole prized innovation. His motto summed it up: "Kick the hell out of the status quo!" Second, Cole's automotive philosophy placed value and emphasis on removing unnecessary weight from a car. Cole actually held that view ahead of its time. He found himself bucking a trend in the early 1950s, because in those days heavy cars were somehow considered "better" and "more luxurious" by virtue of sheer mass and volume. Lightness and efficiency hadn't yet become the engineering watchwords they are today.

Third, Cole valued performance. He was himself a hard, fast driver, and he believed cars ought to move when he wanted them to. He liked cars that performed and disliked those that didn't.

And fourth, Cole didn't mind spending money when the end justified the means. He had the conviction, the enthusiasm, the technical knowledge, and the speaking ability to make even General Motors' most pinchpenny accountants come around to his way of thinking. That's one talent Cole brought to bear again and again in the 1955 Chevrolet's conception.

Chevrolet was working on a "new" 1955 line even before Ed Cole got there in May of 1952. An entirely different car was being developed under Cole's predecessor, a man named Edward H. Kelley and known fondly by his people as "Crankshaft" Kelley.

Kelley preceded Cole as Chevrolet's chief engineer and, while in that position, he was certainly competent. But Kelley was also a man from an earlier era. He'd gotten his nickname by designing the crankshaft for the original Stovebolt Six.

This became the one-and-only engine that, with minor changes, powered every Chevrolet car and truck from 1929 through 1954. To Crankshaft Kelley, the in-line, overhead-valve Six spelled *Chevrolet*. He saw it as America's engine of choice. But although Kelley wasn't ready to admit it, the Stovebolt's days were numbered. The handwriting on the wall said, "Ford V8," "Oldsmobile V8," "Cadillac V8," "Studebaker V8," and for 1953, which was clearly visible in General Motors' 1951 crystal ball, "Buick V8."

What was also clear was that all these V8-engined cars were aimed at, and appealed to, a more youthful, more performance-oriented buyer. That buyer was turning more and more to Ford and, by 1954, with Ford's newly created overhead-valve V8, Ford actually held 47% of the low-priced field, a figure that worried General Motors more than a little.

It worried GM enough, in fact, that the corporation decided that Chevrolet should not only go ahead seriously on a V8 but, beyond that, that the division needed to think younger. That's *the* reason General Motors gave its approval to the Corvette for 1953. Because GM management, to their everlasting credit, could see that Chevrolet—despite its tremendous financial success—was losing ground and becoming an old man's car. Ford V8s appealed to kids—hot rodders—while Chevrolets were being bought and driven by those kids' granddads.

Chevrolet had slowly, almost imperceptibly been veering to the right. The look, the feel, the design and engineering of Chevy's 1953-54 models confirmed the division's conservatism. Looking at those cars before they went into production, the men on GM's 14th floor couldn't help but ask themselves: Should we continue in the 1953-54 direction? Chevrolet had to change. The make had to remake itself. Chevrolet needed to move beyond the Six and modernize with a V8. It needed more sizzle from stop lights and more pizzaz in its styling so that even sedans and wagons would appeal to younger buyers.

It was also clear to GM and Chevrolet executives that Ed Kelley was not the man to make the division more youthful. Kelley reluctantly took on the V8 assignment but didn't have his heart in it. He loved the Six too much. His background was also in manufacturing, and he held all the old manufacturing tenets that too often said, "It can't be done."

Ed Cole made his initial reputation at Cadillac by spearheading development of that division's 1949 overhead-valve V8, an engine as revolutionary in its way as the Chevrolet V8 would become in 1955. And for many of the same reasons: It was lightweight, short-stroke, high-revving, fuel-efficient, and powerful. For 1949, the Cadillac V8 represented state of the art.

But even as early as 1951, that engine's superiority was challenged by Chrysler, which put its "hemi" into production that year. And while the hemi was a relatively large, heavy, expensive V8, it, too, represented a bold step in state of the art. Hemispherical combustion chambers and canted overhead valves weren't new, but Chrysler managed to make that technology available at or just below Cadillac's price range.

In late 1951 and early 1952, Ed Cole had been in charge of GM's military tank plant in Cleveland, Ohio. The Korean war was raging, and the Cleveland facility had been pressed into service to quickly turn out large numbers

One of Chevrolet's rare engineering conferences took place in mid-April of 1953. Participants were (from right to left): Proving Grounds road test supervisor Frank Torrance, chief experimental engineer Rosey Rosenberger, assistant chief engineer Harry Barr, chassis design engineer Russ Sanders, truck chassis engineer Bob Plexico, director of research and development Maurice Olley, plant engineer Mac McCullough, truck chassis engineer Ed Sprung, body engineer Lux Luxmoore, chief engineer Ed Cole, (an unidentified participant), assistant chief body engineer Jim Premo, chief production engineer Bill Flowers, production engineer Walt Frederick, administrative engineers Art Altz and By Holmes, and engineer in charge of Chevrolet's test laboratories Art Underwood.

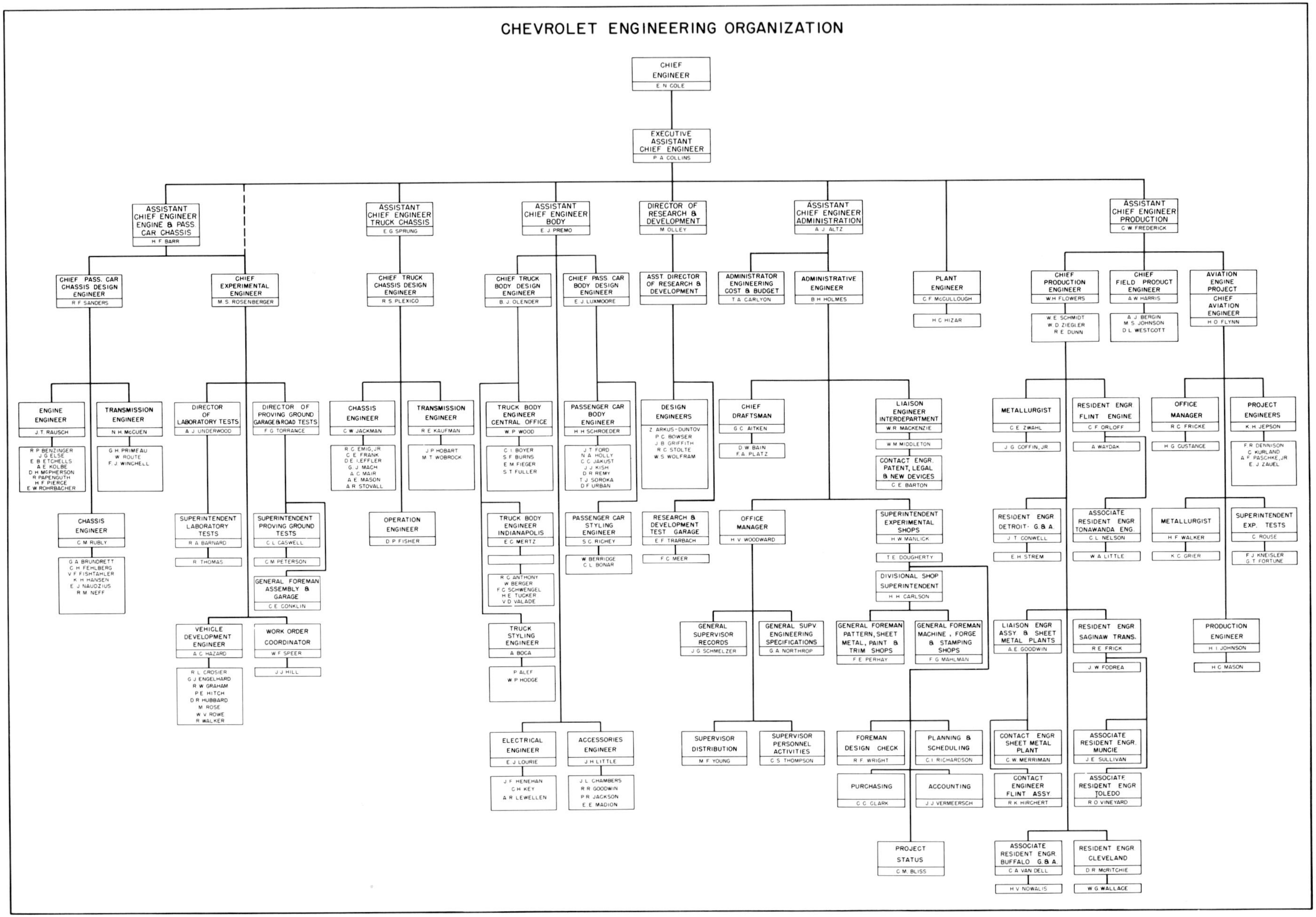
CHEVROLET ENGINEERING ORGANIZATION
CHIEF ENGINEER
E N COLE
EXECUTIVE ASSISTANT CHIEF ENGINEER
P A COLLINS
ASSISTANT CHIEF ENGINEER ENGINE & PASS. CAR CHASSIS
H F BARR
CHIEF PASS. CAR CHASSIS DESIGN ENGINEER
R F SANDERS
CHIEF EXPERIMENTAL ENGINEER
M. S. ROSENBERGER
ENGINE ENGINEER
J. T. RAUSCH
R P BENZINGER
J G ELSE
E B ETCHELLS
A E KOLBE
D H MCPHERSON
R PAPENGUTH
H F PIERCE
E W ROHRBACHER
TRANSMISSION ENGINEER
N. H. McCUEN
G H PRIMEAU
W ROUTE
F. J. WINCHELL
CHASSIS ENGINEER
C M RUBLY
G A BRUNDRETT
C H FEHLBERG
V F FISHTAHLER
K H HANSEN
E J NAUDZIUS
R M NEFF
DIRECTOR OF LABORATORY TESTS
A J UNDERWOOD
SUPERINTENDENT LABORATORY TESTS
R A BARNARD
R THOMAS
DIRECTOR OF PROVING GROUND GARAGE & ROAD TESTS
F G TORRANCE
SUPERINTENDENT PROVING GROUND TESTS
C L CASWELL
C M PETERSON
GENERAL FOREMAN ASSEMBLY & GARAGE
C E CONKLIN
VEHICLE DEVELOPMENT ENGINEER
A C HAZARD
R L CROSIER
G J ENGELHARD
R W GRAHAM
P E HITCH
D R HUBBARD
M ROSE
W V ROWE
R WALKER
WORK ORDER COORDINATOR
W F SPEER
J J HILL
ASSISTANT CHIEF ENGINEER TRUCK CHASSIS
E G SPRUNG
CHIEF TRUCK CHASSIS DESIGN ENGINEER
R S PLEXICO
CHASSIS ENGINEER
C W JACKMAN
R C EMIG, JR
C E FRANK
D E LEFFLER
G J MACH
A C MAIR
A E MASON
A R STOVALL
TRANSMISSION ENGINEER
R E KAUFMAN
J P HOBART
M T WOBROCK
OPERATION ENGINEER
D P FISHER
ASSISTANT CHIEF ENGINEER BODY
E J PREMO
CHIEF TRUCK BODY DESIGN ENGINEER
B J OLENDER
TRUCK BODY ENGINEER CENTRAL OFFICE
W P WOOD
C I BOYER
S F BURNS
E M FIEGER
S T FULLER
TRUCK BODY ENGINEER INDIANAPOLIS
E C MERTZ
R C ANTHONY
W BERGER
F C SCHWENGEL
H E TUCKER
V D VALADE
TRUCK STYLING ENGINEER
A BOCA
P ALEF
W P HODGE
CHIEF PASS CAR BODY DESIGN ENGINEER
E J LUXMOORE
PASSENGER CAR BODY ENGINEER
H H SCHROEDER
J T FORD
N A HOLLY
C C JAKUST
J J KISH
D R REMY
T J SOROKA
D F URBAN
PASSENGER CAR STYLING ENGINEER
S C RICHEY
W BERRIDGE
C L BONAR
ELECTRICAL ENGINEER
E J LOURIE
J F HENEHAN
C H KEY
A R LEWELLEN
ACCESSORIES ENGINEER
J H LITTLE
J L CHAMBERS
R R GOODWIN
P R JACKSON
E E MADION
DIRECTOR OF RESEARCH & DEVELOPMENT
M OLLEY
ASST. DIRECTOR OF RESEARCH & DEVELOPMENT
DESIGN ENGINEERS
Z ARKUS-DUNTOV
P C BOWSER
J B GRIFFITH
R C STOLTE
W S WOLFRAM
RESEARCH & DEVELOPMENT TEST GARAGE
E F TRARBACH
F C MEER
ASSISTANT CHIEF ENGINEER ADMINISTRATION
A J ALTZ
ADMINISTRATOR ENGINEERING COST & BUDGET
T A CARLYON
ADMINISTRATIVE ENGINEER
B H HOLMES
CHIEF DRAFTSMAN
G C AITKEN
D W BAIN
F A PLATZ
LIAISON ENGINEER INTERDEPARTMENT
W R MACKENZIE
W M MIDDLETON
CONTACT ENGR. PATENT, LEGAL & NEW DEVICES
C E BARTON
OFFICE MANAGER
H V WOODWARD
GENERAL SUPERVISOR RECORDS
J G SCHMELZER
GENERAL SUPV. ENGINEERING SPECIFICATIONS
G A NORTHROP
SUPERVISOR DISTRIBUTION
M F YOUNG
SUPERVISOR PERSONNEL ACTIVITIES
C S THOMPSON
SUPERINTENDENT EXPERIMENTAL SHOPS
H W MANLICK
T E DOUGHERTY
DIVISIONAL SHOP SUPERINTENDENT
H H CARLSON
GENERAL FOREMAN PATTERN, SHEET METAL, PAINT & TRIM SHOPS
F E PERHAY
GENERAL FOREMAN MACHINE, FORGE & STAMPING SHOPS
F G MAHLMAN
FOREMAN DESIGN CHECK
R F WRIGHT
PLANNING & SCHEDULING
C I RICHARDSON
PURCHASING
C C CLARK
ACCOUNTING
J J VERMEERSCH
PROJECT STATUS
C M BLISS
PLANT ENGINEER
C F McCULLOUGH
H C HIZAR
ASSISTANT CHIEF ENGINEER PRODUCTION
C W FREDERICK
CHIEF PRODUCTION ENGINEER
W H FLOWERS
W E SCHMIDT
W D ZIEGLER
R E DUNN
CHIEF FIELD PRODUCT ENGINEER
A W HARRIS
A J BERGIN
M S JOHNSON
D L WESTCOTT
AVIATION ENGINE PROJECT
CHIEF AVIATION ENGINEER
H O FLYNN
METALLURGIST
C E ZWAHL
J G COFFIN, JR
RESIDENT ENGR. FLINT ENGINE
C F ORLOFF
A WAYDAK
RESIDENT ENGR DETROIT- G & A.
J T CONWELL
E H STREM
ASSOCIATE RESIDENT ENGR TONAWANDA ENG.
C L NELSON
W A LITTLE
LIAISON ENGR ASSY. & SHEET METAL PLANTS
A E GOODWIN
CONTACT ENGR SHEET METAL PLANT
C W MERRIMAN
CONTACT ENGINEER FLINT ASSY.
R K HIRCHERT
RESIDENT ENGR. SAGINAW TRANS.
R E FRICK
J W FODREA
ASSOCIATE RESIDENT ENGR. MUNCIE
J E SULLIVAN
ASSOCIATE RESIDENT ENGR TOLEDO
R O VINEYARD
ASSOCIATE RESIDENT ENGR. BUFFALO G & A.
C A VAN DELL
H V NOWALIS
RESIDENT ENGR. CLEVELAND
D R McRITCHIE
W G WALLACE
OFFICE MANAGER
R C FRICKE
H G CUSTANCE
PROJECT ENGINEERS
K H JEPSON
F R DENNISON
C KURLAND
A F PASCHKE, JR
E J ZAUEL
METALLURGIST
H F WALKER
K C GRIER
SUPERINTENDENT EXP. TESTS
C ROUSE
F J KNEISLER
G T FORTUNE
PRODUCTION ENGINEER
H I JOHNSON
H C MASON

of tanks. Cole oversaw production of those tanks and did a good job. The Cleveland plant's efficiency, along with Cole's work on the 1949 Cadillac V8, impressed GM management. They reasoned that this might be just the man to turn Chevrolet around; the sort of young blood Chevy needed.

So they summoned Cole to Detroit for an interview. The committee that interviewed Ed Cole consisted of three of the highest-ranking men in the corporation: GM president Charles E. Wilson, executive vice president Harlow H. Curtice, and GM's vice president of operations, Louis C. Goad.

Together they explained the Chevrolet situation to Cole, mentioning, among other things, that Ed Kelley was struggling to design a new low-cost V8. During this interview, the three executives expressed the hope that the 1955 Chevrolet could break with the past and would somehow become more youthful. They weren't quite sure how that might happen, but perhaps Cole could help.

Cole asked some pointed questions about Kelley's V8 and learned it would probably be built along the lines of the 1949 Cadillac and Oldsmobile V8s. That would make it expensive, observed Cole. He also expressed the opinion that there were ways to engineer a less costly V8—new technologies—and he said that Chevrolet ought to explore those possibilities.

Based on the discussion to this point, the committee asked Cole if he'd consider becoming Chevrolet's manufacturing manager. The idea was for Cole to work with Kelley on an equal footing. Neither would be the other's boss, but together they could help each other produce a Ford-beating, V8-engined 1955 Chevrolet. Wilson, Curtice, and Goad also let it be known that if Cole would join Chevrolet in 1952, he could count on them to consider making him the division's general manager at some time in the future.

Cole quickly thought this offer over and then slowly, with that schoolboy smile that was so much a part of his persuasiveness, suggested another possibility. He told the committee that he felt he could be of more service to Ed Kelley and of more value to General Motors if they made Kelley manufacturing manager and Cole chief engineer. Cole argued that otherwise he, Cole, as manufacturing manager, could only produce what Kelley and Chevrolet Engineering designed. As chief engineer, Cole could put forth his ideas for a new, lightweight, less expensive V8 that should do everything Chevrolet might hope for in terms of youth appeal. He acknowledged that Ed Kelley had tremendous manufacturing ability. With Kelley's approval, the committee should ask him to make a horizontal move and become Chevrolet's manufacturing manager, and Cole would work with him to conceive a car to bring Chevrolet out of the doldrums.

Those words capture the essence of Cole's remarks. I don't believe records of that meeting exist; I haven't been able to find them. But the details have come out in some of my interviews, particularly those with Cole himself and with Walt Mackenzie. The upshot was that Ed Kelley very graciously accepted Cole's proposal. And since the positions of Chevrolet chief engineer and manufacturing manager really *were* equal in rank, it wasn't a step down for Kelley.

I don't want give the impression that the 1955 Chevrolet sprang full-blown from Cole's mind; that he did everything all by himself. While it's true that he was a dynamo and the prime mover behind the '55's development,

These are the only known photos of the Chevrolet V8 developed under Ed Kelley before Cole arrived. This engine was modeled on the 1949 Cadillac/Olds V8s, with thickwall castings, shaft rockers, and complex intake manifolds. White lines in upper photo are shrink cracks in original negative.

he needed quite a large staff to bring the car into being. And he quickly assembled that staff.

When Cole took over Kelley's office in May 1952, he immediately did several important things. Within a few months, he boosted Chevrolet's engineering pool from 850 to nearly 3000 people. He did this by "putting out the call." He specifically invited the best and brightest of GM's young engineers to join him at Chevrolet. In addition, volunteers came to him from every direction. Everyone knew Cole's reputation—his enthusiasm, that he was a wellspring of good ideas, dealt fairly with his people, and that he was generally great fun to work for.

Cole began work on the '55 passenger car and trucks by summarily tossing out virtually every previous program that Chevrolet Engineering had set up. The notable exception was the Corvette. The 1955 line started, in other words, afresh, with a clean sheet of paper. And that also appealed to the people working with him: to create an all-new, clean-sheet car. But the program included more than just a new car. It also called for an entirely new line of pickups and heavy trucks, and there was as much work involved in developing those as for the passenger car.

Cole put Harry Barr in overall charge of the 1955 V8 engine and the passenger-car chassis. "Chassis" meant frame, suspension, and running gear. Ellis J. (Jim) Premo had responsibility for body development: passenger-car *and* the truck bodies. Premo's became a 2-pronged job that involved not only the creation of these bodies' cowl structures, floorpans, and outer skins but also liaison with Fisher Body Div., which would do the actual manufacturing. Passenger-car bodies were shared, of course, with Pontiac.

Maurice Olley became Cole's director of research and development. Olley, one of the world's foremost suspension engineers, had come to General Motors from Rolls-Royce in 1930 and had been instrumental in developing GM's independent A-arm front suspension for 1934. Now, as Chevrolet's R&D director, Olley took a number of young design engineers under his wing, including Zora Arkus-Duntov, who would soon inherit the Corvette program.

The job of engineering the 1955 Chevrolet would take roughly 24 months and would be complicated by the fact that Cole wanted to use a number of new engine innovations that had never been tried before in mass production. One was the stamped rocker arm. The stamped rocker arm was not an idea that originated at Chevrolet; it was invented by Pontiac engineer Clayton B. Leach in his basement workshop at home.

FUTURE ENGINES

NEW BASIC DESIGN IS REQUIRED FOR FURTHER INCREASES IN OUTPUT AND FUEL ECONOMY.

INVESTIGATION OF NEW DESIGNS HAS LED TO CONCENTRATION OF DEVELOPMENT WORK ON A NEW IN-LINE SIX AND A V-8.

BOTH TYPES ARE BEING PREPARED SINCE EACH HAS ADVANTAGES WHICH CHEVROLET WANTS . . .

IN-LINE SIX	V-8
○ HIGHER SPECIFIC OUTPUT	○ SMOOTHER, QUIETER
○ LIGHTER	○ SHORTER
○ FEWER PARTS, LESS FRICTION	○ STRUCTURALLY MORE RIGID
○ NARROWER, LESS CROWDED INSTALLATION	
○ LESS COSTLY	

BOTH ARE DESIGNED TO OPERATE EFFICIENTLY ON PRESENT FUELS AND CAPABLE OF COMPRESSION RAT:O INCREASES TO UTILIZE FUELS OF THE FORESEEABLE FUTURE.

FUTURE ENGINES
COMPARATIVE SPECIFICATIONS

		IN-LINE SIX	V-8
SIZE RANGE	BORE	3-15/16 TO 4-1/8	3-1/2 TO 3-15/16
	STROKE	3-1/4 TO 3-3/4	3
	DISPLACEMENT	230 TO 301	231 TO 292
COMPRESSION RATIO RANGE		8-TO-1 TO 12-TO-1	
ESTIMATED WEIGHT (NOMINAL 260 CU. IN. SIZE)		550	
NUMBER OF MAIN BEARINGS		7	5
VALVE LOCATION		IN CYLINDER HEAD	
PISTON MATERIAL		ALUMINUM	
LUBRICATION SYSTEM		FULL PRESSURE	

ENGINE PERFORMANCE COMPARISON

FUTURE 256 CU. IN. DISPLACEMENT V-8 8.0-TO-1 C.R.
VS.
1953 235 CU. IN. DISPLACEMENT SIX 7.75-TO-1 C.R.

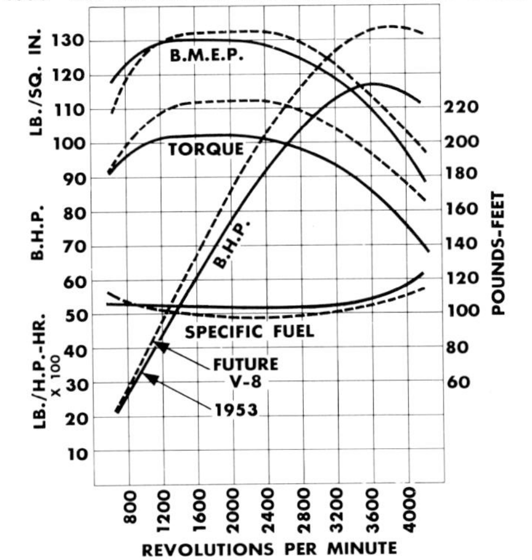

As soon as they arrived at Chevrolet in May 1952, Ed Cole and his staff began outlining targets and goals for the 1955 V8. The V8's initial displacement was projected at 256 cubic inches, but future increases ranged up to 292. Compression ratios ran as high as 12:1.

1955 PROPOSED WEIGHT REDUCTION

4 DOOR SEDAN

	ESTIMATED WEIGHT - POUNDS	
	1953	1955
BODY AS PURCHASED	1072	1040
BODY PARTS FURNISHED BY CHEVROLET	29	20
BODY MOUNTINGS	21	20
CHASSIS SHEET METAL	142	132
RADIATOR GRILLE AND RADIATOR SUPPORTS	31	31
BUMPERS, BUMPER GUARDS, AND REAR GRAVEL DEFLECTOR	94	94
BODY AND SHEET METAL TOTAL	1389	1337
SIX CYLINDER CHASSIS TOTAL	1812	1713
COMPLETE VEHICLE SHIPPING WEIGHT	3201	3050

WEIGHT REDUCTION
BODY AND SHEET METAL—52 POUNDS
COMPLETE VEHICLE—151 POUNDS

BODY AND SHEET METAL INTERCHANGEABILITY

PASSENGER CAR MAJOR PANELS

REDUCED TOOLING AND MANUFACTURING COSTS MAY BE OBTAINED THROUGH GREATER INTERCHANGEABILITY OF MAJOR BODY AND SHEET METAL PANELS.

IN 1953, 15 PASSENGER CAR MODELS AND SEDAN DELIVERY REQUIRE 144 MAJOR PANELS.

FOR FUTURE MODELS, IT IS PLANNED TO REDUCE THE NUMBER OF DIFFERENT PANELS BY REVISED STYLING WHICH WOULD PERMIT GREATER INTERCHANGEABILITY. A REDUCTION OF 24 PANELS IS PLANNED.

PROPOSED PANEL CHANGES

ITEM	NUMBER OF DIFFERENT PANELS		REDUCTION
	1953	FUTURE	
ROOF PANEL	7	4	3
WINDSHIELD INNERFRAME	3	2	1
CENTER PILLAR (L. H. & R. H.)	4	2	2
REAR QUARTER OUTER PANEL (L. H. & R. H.)	16	*14	2
REAR QUARTER INNER PANEL (L. H. & R. H.)	14	*16	2 (ADDED)
DECK LID OUTER	2	1	1
DECK LID INNER	2	1	1
FRONT DOOR OUTER (L. H. & R. H.)	8	6	2
REAR DOOR OUTER (L. H. & R. H.)	4	2	2
FRONT DOOR INNER (L. H. & R. H.)	8	6	2
REAR DOOR INNER (L. H. & R. H.)	4	2	2
PACKAGE TRAY	4	2	2
LOWER BACK PANEL	3	2	1
FRONT END SIDE RAIL (L. H. & R. H.)	6	4	2
HOOD PANEL (L. H. & R. H.)	2	1	1
FRONT FENDER BAFFLE (L. H. & R. H.)	2	0	2
NET REDUCTION			24

*INCLUDES TWO PANELS REQUIRED FOR 2-DOOR STATION WAGON

Reductions in body weight and complexity were important parts of Cole's 1955 program. Body engineer Jim Premo managed to get the number of panels down from 144 for the 1953 Chevrolet to 120 for 1955. The 1955 4-door sedan ended up being 151 pounds lighter.

To make Leach's stamped rocker arm practical, Barr's engine group had to develop and test not only the rocker itself but the pressed-in mounting stud on which it stood, the hollow pushrod with hardened ends that fed oil to the engine's topside, and the top-oil metering system—the little "wafers" developed by Loren R. Papenguth and incorporated into the hydraulic valve lifters. Without Papenguth's wafers, the engine's upstairs would be awash in excess oil. None of that technology existed at the beginning of the Chevrolet V8 program.

Also untried were the new processes for casting the 1955 block and heads. The idea was to make the Chevy V8's block and heads as light and strong as possible by using what today we call "thinwall" casting techniques. This meant that the castings had to be dead-on accurate. Precision resulted in less machining, but it also took away the luxury of error margins, because you now couldn't machine to compensate for core shift. Core shift is the movement of the permanent cores when metal is poured into the mold; common in the days of thickwall castings.

What Cole and Barr (and Kelley) were counting on was a casting process developed by John Dolza at GM central Engineering Staff. Dolza called it his "green sand" casting process, and for this he used a slightly wet core sand, which allowed casting the block upside down with fewer permanent cores than were used traditionally. The Cadillac V8 of that era used 22 permanent cores for the block alone, all held in place by steel rods. The more cores and rods, Dolza argued, the more chances for the cores to shift

While final configurations weren't yet set, wooden mockups gave Chevrolet engineers and stylists some idea of the spatial dimensions of the new V8 and inner body panels. Early mockups included a front sway bar, which appeared on taxis and police cars only.

These craftsmen, creators of the many wooden '55 components, were mostly older men. They would soon be superseded by computerized milling machines. For now, though, there was virtually nothing they couldn't replicate in wood, plaster, fiberglass, and sheetmetal.

Swivel bracket at right was considered for power-steering pump, with air-conditioning compressor up top and generator at left.

Twin steering boxes grace left mockup, one for export. Power steering pump did mount to generator but not as on this wooden rendition.

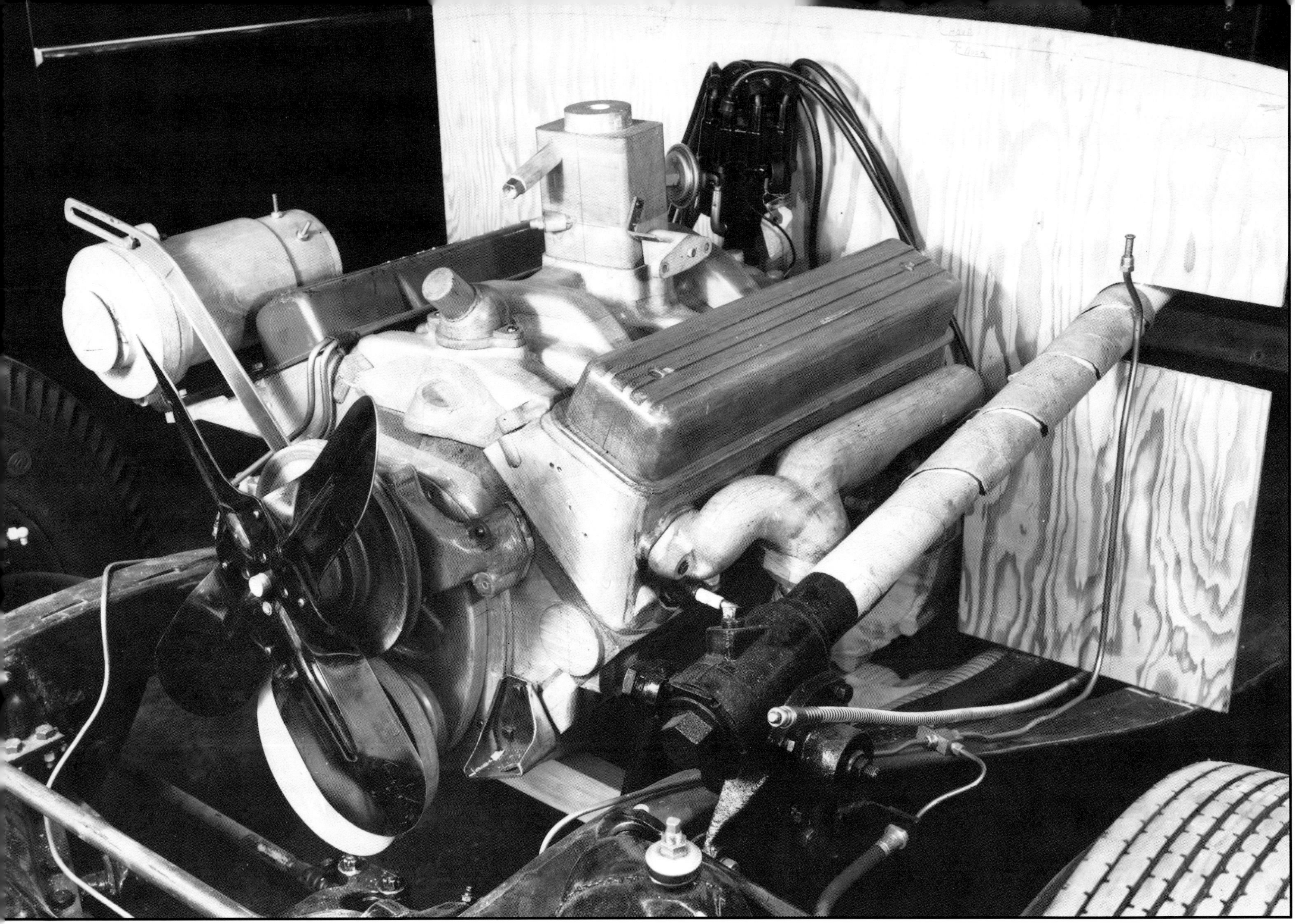

It's interesting to see how the rocker covers and other V8 details changed over the course of the 1955 development program.

Air conditioning was envisioned as a drop-in. Bendix Treadle Vac power brake booster, shown, became available on '55 Powerglide cars.

Bendix Hydrovac, shown above, became available for 1955 trucks and 1957 cars with stick shift. This mockup has 6-cylinder radiator and vacuum wiper motor. For production 1955 cars with air, firewall had to have special holes made with an air chisel, trim tag moved right.

when molten metal was poured in. And to compensate for core shift, cylinder walls and decks had to be cast thicker, then later corrected by machining.

Dolza, working alongside Harry Barr's engine group under John T. Rausch, managed to pare the number of permanent cores for the V8 from 22 to 12. Together, they also eliminated all core-supporting steel rods. The amount of sand used to cast each '55 Chevrolet V8 was 131 pounds, versus 222 pounds for the Cadillac V8. Foundry precision with the green-sand process was so good that wall thicknesses could be cut nearly in half. The metal saved, combined with fewer machining operations, eventually helped make the Chevrolet V8 as economical to manufacture as the in-line Six. And the small-block V8 weighed 41 pounds less than the Stovebolt (531 versus 572 pounds).

One of the V8 engine's givens *was* light weight. To make it light but rigid, the block and crankshaft had to be kept as short and compact as possible. Ed Kelley's initial V8 was planned with a displacement of 231 cubic inches, which Cole and Barr immediately recognized as too small. They soon drew up a 245-cid V8 but decided they could make it 260 with no weight penalty. After considerable discussion, they settled on a 3.75-inch bore, which meant that 260 cubes would have dictated a stroke of 2.93

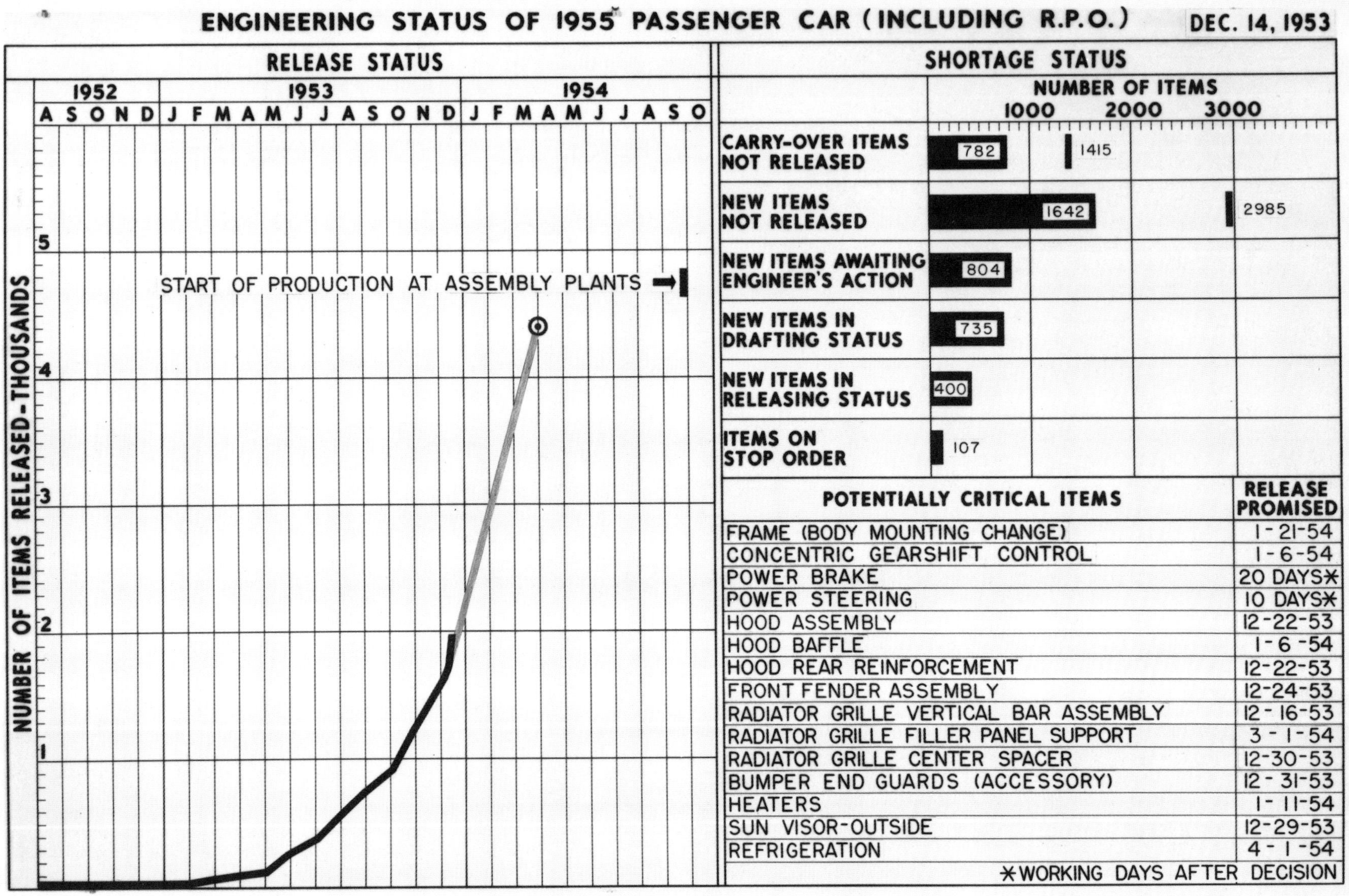

At the time of the 1955 Chevrolet's development, specific components were "released" or approved for production as they were successfully engineered and tested. This chart shows engineering releases starting in Jan. '53, with final production and assembly slated for Sept. 1954. Today, so-called "simultaneous engineering" allows all or most parts of a new design to be released at the same time.

Early V8 casting, dated 5/53, shows significant detail differences from production, including fuel-pump pad, manifolds. Heads are stamped "dummy," which means there weren't any passages inside. This engine is shown in a truck chassis, as evidenced by leaf springs.

inches. Barr decided to round this off to 3.0 inches even, resulting in the 265 cubic inches that the 1955 Chevrolet V8 had at birth.

One odd point: Early prototypes of Cole's V8 had the two exhaust manifolds cast integrally with the crossover pipe. In other words, the exhaust system was all one piece: manifolds and crossover cast as a single unit. What the advantage might have been isn't clear; perhaps a cost saving. In any case, the experiment failed, because different cooling rates along those several feet of cast-iron pipework caused the crossover to crack. The idea was soon abandoned. The conventional steel-tube crossover turned out to be lighter and more forgiving.

The intake manifold for this engine became a single casting that took on several duties in addition to distributing fuel. The intake manifold served as the engine valley cover and eliminated the need for a conventional pressed sheetmetal cover. It housed the oil filler tube, provided water outlets to the radiator and heater, housed the cooling-system thermostat and carburetor heat stove, and had a place machined into it to mount the ignition distributor. The intake manifold was also cast in such a way that its parting seams were along the bottom, where flashing wouldn't interfere with seating against the heads.

Among this V8's further sophistications, the engine needed no drilled oil

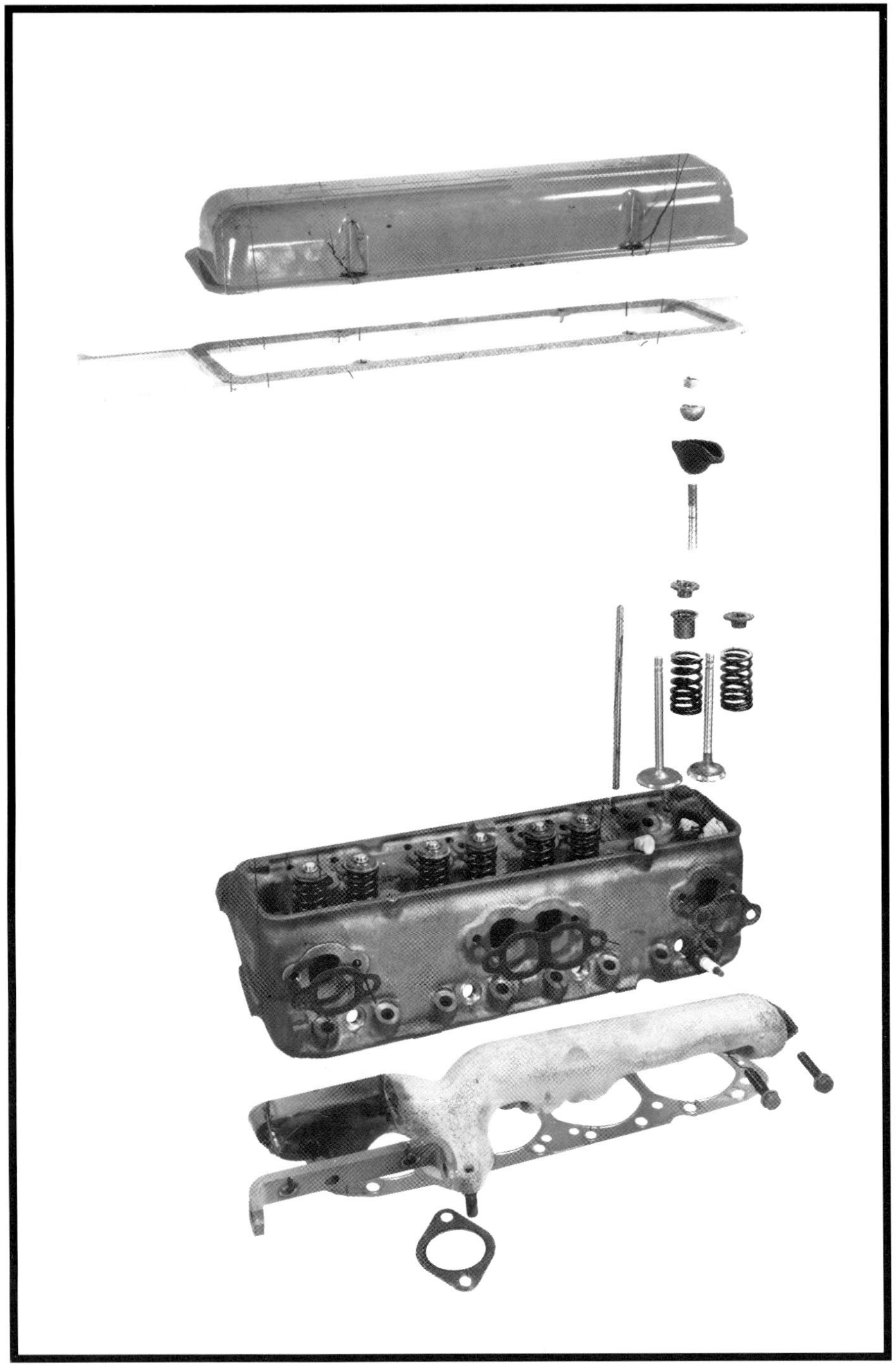

Exploded view of 1955 production small-block V8 shows relationships of components. Features that distinguished this engine from previous

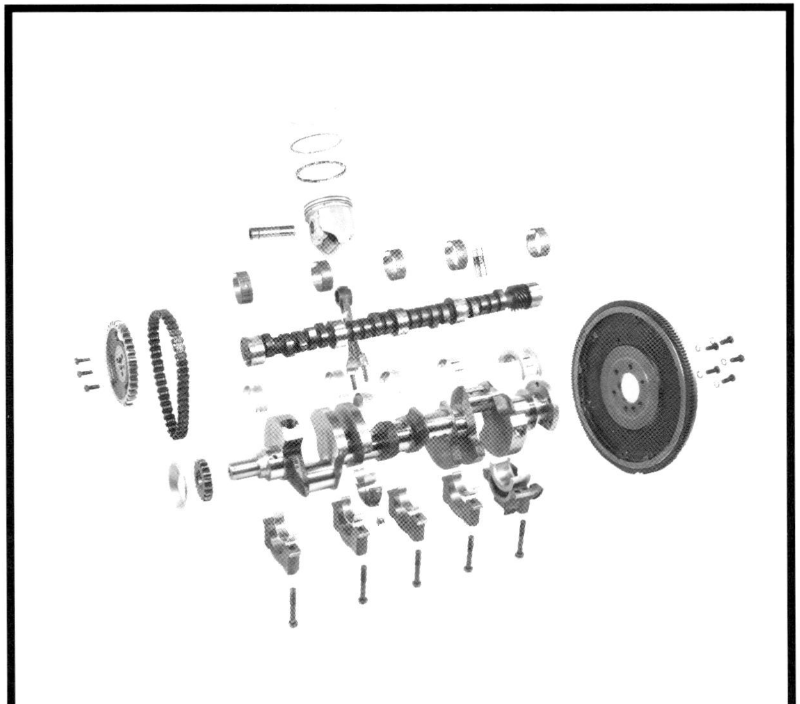

V8s were thinwall castings to keep weight down and stamped rockers instead of shafts, which made each valve independent.

Another unusual feature of new V8 was its intake manifold, which doubled as the engine's valley cover. With its casting seams at the bottom, it needed little machining. Early 1955 swing-type oil pump pickup was soon eliminated due to cost.

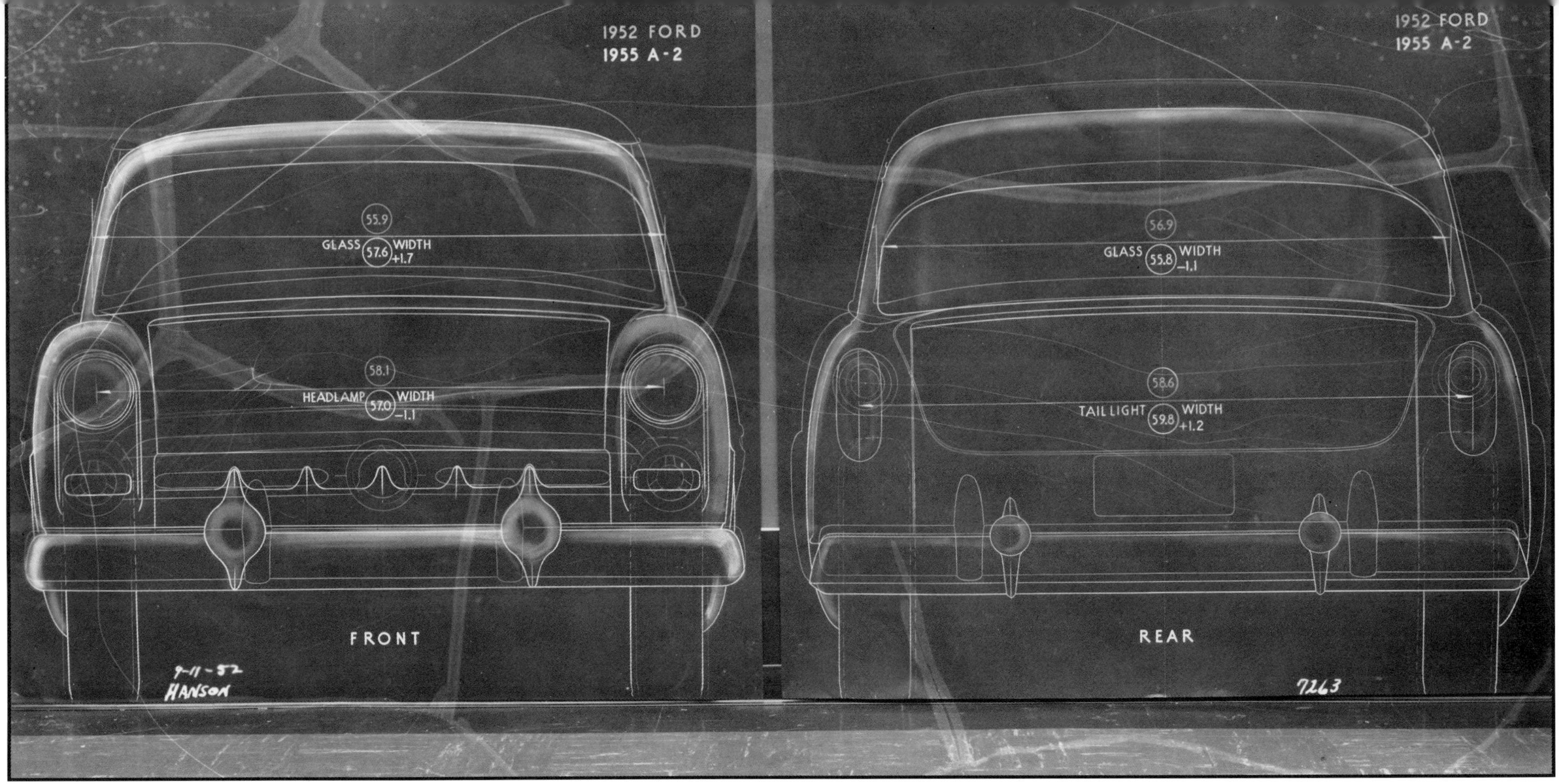

Chuck Stebbins' full-sized body room drawings, dated 9/52, show proposed low-hood 1955 Chevrolet overlaid against 1952 Ford.

galleries above the deck. Valve lubrication was supplied, as mentioned, through the pushrods. The waterpump became a simple, compact scroll design.

The 1955 Chevrolet's body began to take shape as early as 1951 in the body development room of what was then called General Motors Styling (in 1972, the name was changed to GM Design Staff). At that time, Styling was very much the domain of Harley Earl, the man who carved his own empire out of raw clay and, in doing so, taught the world's major auto companies how to design cars. Harley Earl made up the rules as he went along; something he had to do because they didn't exist before.

A tall, imperious, unpredictable gentleman with a sometimes sunny and often fierce disposition, Earl ruled with the assured backing of GM board chairman Alfred P. Sloan Jr. When any of GM's divisional chiefs or other vice presidents questioned Earl's authority, Earl would simply pick up the telephone and say, "Well, let's see what Mr. Sloan thinks about that," which promptly ended the discussion. Earl's tremendous authority left some of his own underlings shaken for life. Harley Earl retired in 1958, but in the early 1950s, when the 1955 Chevrolet was being designed, he was at the peak of his powers, both aesthetically and dictatorially.

In those days, you'll remember, four of the five GM car lines had their own distinctive recognition symbols. Tailfins and a golden V meant Cadillac, portholes said Buick, Oldsmobile used the rocket and globe, and Pontiac had its twin silver streaks. Those weren't the only identifiers, but they were primary and longterm.

The only GM division without a styling hallmark was Chevrolet. Harley Earl, who overlooked nothing, had consciously decided not to give Chevrolet a hard-and-fast identifier. Chevy styling purposely stayed flexible, partly so that Chevrolet could, in some years, take on symbols or attributes from other car lines. For example, Chevrolets looked like miniature Cadillacs in model-years 1932 and 1939 and like smaller versions of Buicks in 1940 and 1941.

The 1955 Chevrolet took shape initially within GM Styling in what was then called "the body room." There were actually three body rooms at that time, one for the A Body (shared by Chevrolet and Pontiac), one for the B Body (Olds and smaller Buicks), and one for the C Body (big Buicks and Cadillac).

The "A" body room was run by Charles A. (Chuck) Stebbins, and it was Stebbins' job to take "the package"— that set of dimensions established jointly by Chevrolet Engineering and Fisher Body—and translate them into scale drawings and preliminary design sketches which Chevrolet stylists could use to fashion the final look of the production car.

Stebbins' body room was on the 10th floor of GM Styling, which back then took up part of the General Motors Research Building. The Research Building stood on West Milwaukee Avenue, in downtown Detroit, just behind the General Motors Building. The two buildings were linked by a bridge and tunnels. The old Research Building, now called the Argonaut Building, serves a different function today, and GM Design Staff is now housed at the General Motors Technical Center 20 miles north of downtown in the suburb of Warren, Mich.

From late 1951 to early 1952, Stebbins fashioned an overall shape for the 1955 Chevy-to-be, based on Chevrolet Engineering's and Fisher Body's figures and dimensions. He and his small staff of draftsmen, designers, and clay modelers also established seating positions, trunk space, and began to guess at the size and position of such mechanical components as the new suspension systems and the not-yet-developed V8 engine.

Stebbins' body room was in constant communication with the Chevrolet styling studio on the floor above, both through Harley Earl himself, who spent time in both places daily, and through Chevrolet's exterior/interior studio chief, Clare (Mac) MacKichan (pronounced MacKEEkan). Mac had taken over the Chevrolet studio on 2 April 1951. It was under him, always monitored by Harley Earl, that the 1955 Chevrolet's exterior and interior would come together.

Harley Earl respected Ed Cole and the two got along well, which wasn't something you could say for Earl and too many other chief engineers, past or contemporary. Engineers and designers, like oil and water, didn't usually mix. MacKichan respected both men, and together they did an extraordinary job of giving shape and sparkle to the 1955 design.

Cole kept in constant contact with MacKichan's styling studio, and while he didn't kibitz, he was vitally interested in how the car would look. The success of its appearance was just as important to him as getting the mechanical parts right. In his mind, the two aspects were inseparable, each adding value to the other. Cole made sure Engineering never stood in the way of Styling; very unlike the adversarial relationship that traditionally plagues those two functions.

It was decided early on that the 1955 Chevrolet should use a wraparound windshield. GM's production B and C bodies already had this feature, and 1955 would be the A's turn. In the interest of economy and repairability, GM's cost accountants suggested going to a 2-piece, split wraparound design, but Cole, Earl, and MacKichan disagreed and eventually won the day. The 1955 Chevrolet's windshield, of course, came out in one piece.

Early sketches and clays of the 1955 Chevrolet show a hood that ducks down between the front fenders. With the prominent front fenders,

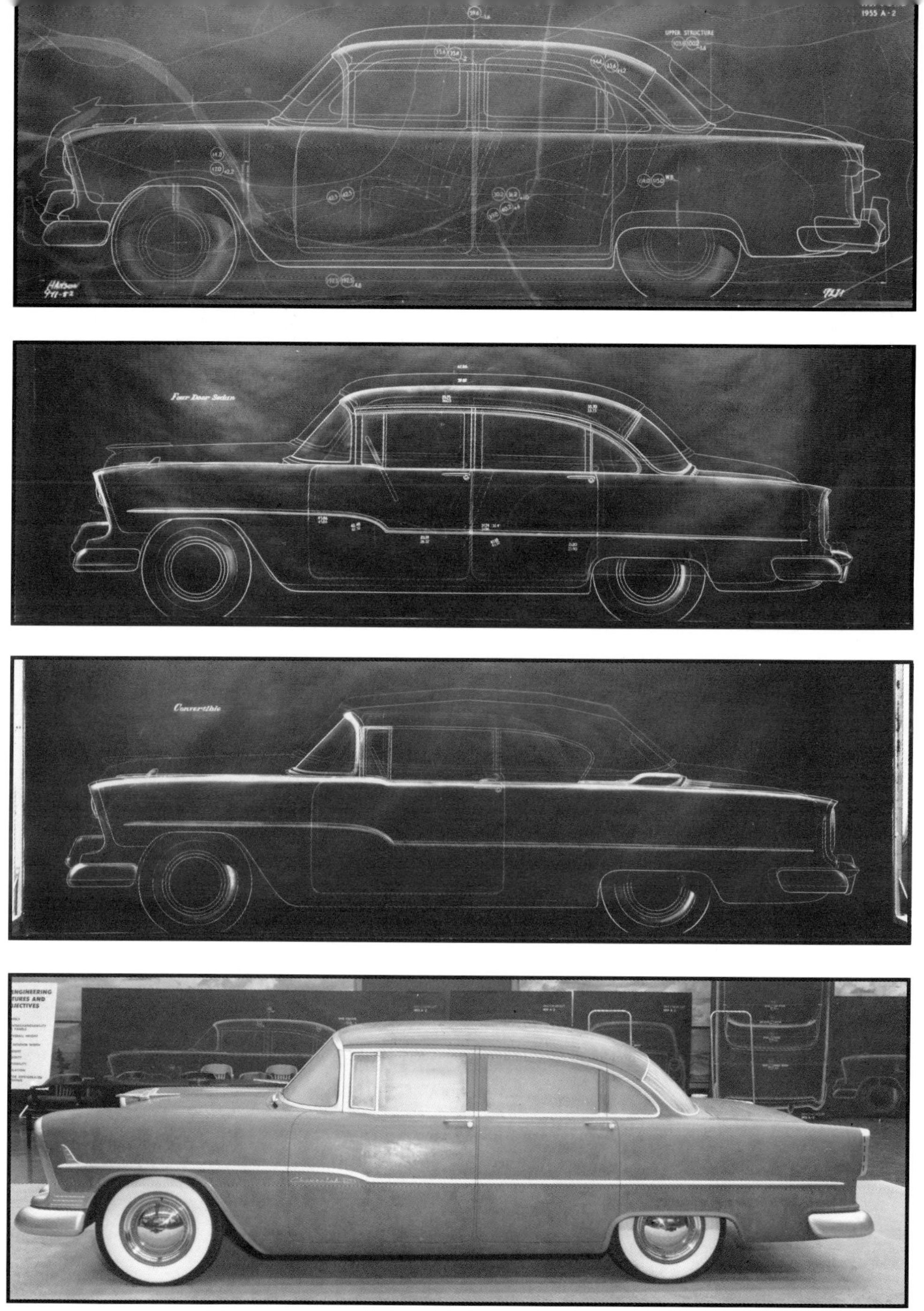

Top drawing shows early Chevy slightly shorter than 1952 Ford and considerably lower than 1953 Chevrolet (second down). Full-sized clay model, with wraparound windshield, was also done in body room.

Very early 1955 scale clay proposal from MacKichan's studio continued the 1953 theme, especially at rear door and conventional windshield.

MacKichan's designers considered putting a drop in the beltline just behind the wraparound windshield pillar. The front fenderline dipped at the A-pillar and then ran straight back along the tops of the doors and back to the tail lights.

MacKichan and Earl pursued the dropped-hood theme for several months. And when the height of the V8 dictated raising the hoodline, the designers also raised the front fenders. Eventually MacKichan realized that the front view was becoming too tall and narrow, so he and Earl abandoned the dropped-hood idea.

After abandoning the recessed hood, the proposed 1955 Chevrolet's front fenders and hood pretty much formed a plateau and, at that point, the dip behind the windshield didn't make sense anymore. Designer Carl Renner then suggested, instead of continuing straight back with the beltline, why not put a dip just ahead of the rear fender, as on the 1954 Buick and Cadillac? This dip near the C pillar came fairly late in the program, and some prototype bodies didn't have it.

Another late arrival was the Ferrari-like grille. Early grille proposals were genetic derivatives of the 1953-54 Chevrolet theme. These, though, were generally thought inappropriate. Cole felt they lacked character, and Harley Earl wasn't pleased either. After a trip to the European motor shows, Earl came back all fired up with the eggcrate grilles he'd seen on Ferraris. That's what we ought to have on this Chevrolet, he told MacKichan. And odd as it sounded at first to place a Ferrari-esque grille in a full-sized American passenger sedan, the idea worked to MacKichan's and Cole's satisfaction.

When the 1955 Chevrolet was finally introduced on Oct. 28, 1954, the car wasn't what a lot of longtime customers expected.* It would take the public a month or so to get used to the new design. Meantime, initial sales were disappointing. After the 1955 Chevrolet had been on sale for several weeks and sales weren't up to expectations, Harlow Curtice called Chevrolet general manager Tom Keating and Cole into his office for a conference. The problem, said Curtice, was that Ferrari grille. People didn't like it: too foreign, too unusual. What Chevrolet needs, he said, is a more conventional grille.

MacKichan's studio was pressed into service and immediately came up with the 1956 grille and front-end design. The 1956 ensemble was briefly considered as a midyear replacement for the 1955 grille. By December, though, 1955 Chevrolets were selling in record numbers, and the crisis ended almost as quickly as it began. The 1956 grille became just that. And, as a quick aside, the 1957 front ensemble was also developed in 1953 in MacKichan's studio as one of the alternatives to the 1955 grille.

In those days, both exterior and interior styling were done in the main Chevrolet studio. There were no separate interior and exterior studios as there are today. There was an interior *soft trim* studio, however, under Edward J. (Ed) Donaldson, whose duties included choosing upholstery materials, patterns, and colors and coordinating these with exterior and

*The 1955 Nomad station wagon and 1955 Corvette were delayed and went on sale as mid-year models on 26 Jan. 1955.

interior paint colors. But the instrument panel, steering column, and interior hardware were all done in MacKichan's studio, mostly under Carl Renner.

Renner had overall responsibility for the interior and especially for the instrument panel. He himself created the deluxe steering wheel and adapted the "double-bubble" panel configuration from the 1953 Corvette, a dashboard design conceived several years earlier by Bob Bartholemew. The tiny bowtie perforations (987 of them, according to Pat Chappell's book, *The Hot One*) in the '55's stainless steel dash insert were also Renner's idea, although the inkwork was done by designer Duane (Sparky) Bohnstedt.

At the same time that GM Styling was working out the 1955 Chevrolet's body details, wooden mechanical mockups and seating bucks were being fabricated in both Engineering and Styling. These mockups became accurate 3-dimensional representations of parts of the finished car, based on the latest but ever-changing 1955 information and specifications. Mockups were most often painstakingly handcarved from wood but sometimes included prototype metal parts or representations made of cardboard, paper, plaster, and/or fiberglass.

The various wooden "components" of the car were put in place so that designers and engineers would know what went where and how things fit together. GM Styling and Chevrolet Engineering each had its own craftsmen who could carve just about anything out of wood. This mockup stage lasted for several months and became an important phase of the design and engineering processes.

Once the basic details of the new engine were worked out, the V8's design changed relatively little between the completion of early running prototypes and production. Much testing, though, remained to be done, both on the engine and the car itself. Some of this testing took place inside Chevrolet's engineering labs, some at the main General Motors Proving Grounds near Milford, Mich., some at GM's newly opened Desert Proving Grounds near Mesa, Ariz., and the rest on public roads around the country.

Pre-prototype cars, called "mules," were cobbled together. They consisted mostly of 1953 and 1954 Chevrolets into which technicians fitted the prototypical 1955 V8 engines and running gear. Every aspect of the 1955 Chevrolet had to be tested and proven out for reliability, efficiency, economy of manufacture, and safety.

The Warner overdrive was something new for Chevrolet and was given quite a lot of over-the-road testing. This was also the first year for Chevrolet's 12-volt electrical system. The main attraction of 12 volts was that it allowed the use of smaller battery cables and finer wire gauges throughout the car. One Chevrolet publication pointed out that every inch of #10 copper wire added $1000 to the expense of producing a model year's worth of cars. That was in 1953 dollars, which today would be many times more due to inflation.

But the 12-volt electrical system held more advantages than just cost reduction. Early experiments had shown that a 6-volt system didn't really have enough voltage to fire the sparkplugs with the new V8's higher compression. Given a .035-inch plug gap, the V8 on a 6-volt system would start misfiring under hard acceleration at around 60 mph in high gear. With the new 12-volt system, even an abnormally wide .050-inch plug gap gave miss-free operation throughout the engine's rev range. Chevrolet engineers

In Aug. 1952, MacKichan's designers came up with this high-fender, low-hood, low-grille design. The V8's height, though, raised hood and fenders, leaving changed proposal looking too tall and narrow.

further noted that 12 volts lent greater generator efficiency and easier starting.

It was soon discovered that the new V8 had such good combustion efficiency and such low heat rejection that the radiator could be made smaller than the Six's. With the same width core, the V8 radiator ended up needing 28 square inches less frontal area than the Six.

A number of mechanical bugs, though, did show up in early testing. One had to do with oil consumption, especially during deceleration when vacuum was high. Part of the problem was high compression coupled with a large bore diameter. Lab tests showed that at high manifold vacuum, the experimental V8 burned about a quart of oil per 200 miles!

Chevrolet engineers found that oil was passing into the combustion chambers because the rings were fluttering up and down in their grooves. The first solution they tried was a stiff, flat spring placed in the second ring groove, directly beneath the ring itself. While this seemed to be working at first, it was soon discovered that the spring was digging itself into the aluminum piston, and that eventually made the oil burning worse than before.

Working with independent piston-ring manufacturers, Chevrolet finally figured out a solution: a circumferential expander for the oil-control (bottom) ring that gave enough axial and radial load to keep oil from getting past it. Not that Chevrolet had fully solved the oil-burning problem. Early production V8s also used oil, as you'll read in some of the interviews. This had to do with the engine plant honing the cylinder walls too smooth, a problem some dealers remedied by pouring Bon Ami down the carburetor!

Additional teething bugs, like squeaky rockers and valves digging into pistons due to mis-installed rocker studs, also showed up in early production, but these were all quickly sorted out and remedied.

In transcribing the following conversations between myself and the people who designed and developed the 1955 Chevrolet, I tried to be as faithful as I could to their words. I edited the transcripts very reluctantly and only when I had to in order to clarify meaning or, in a few rare cases, to make the conversation easier to follow.

The 1955 Chevrolet turned out to be a remarkable car and one that had a dramatic effect on automotive history. As I write this, the car is very much with us, many examples still soldiering faithfully. The same basic engine is being used, of course, in contemporary Corvettes, Camaros, Caprices, and other GM cars and trucks. Ed Cole's vision has followed us into the 1990s, and so have the talents of those thousands of individuals who worked together so skillfully and enthusiastically to create a great American automobile. **— Michael Lamm**

Seating bucks from Sept. 1952 showed an entirely different decor than chosen for production. Of interest, though, are wraparound windshield and dashboard cover, early steering column, and simple horizontal door trim. White streaks are again cracks in negatives.

Chevrolet styling studio personnel posed in front of 1953 Corvette drawing. Staff members were, from upper left to right: (unidentified studio engineer), designer Bob Cadaret, modeler Bob Diebold, chief modeler Jack Park, designer Sparky Bohnstedt, designer Bill Beuchler, designer Don Schumer, assistant chief designer Joe Schemansky, chief designer Clare MacKichan, and chief studio engineer Hans Hierta.

Clare MacKichan

1955 Chevrolet exterior and interior design studio chief.

Clare (Mac) MacKichan had yearned to design automobiles since boyhood. He grew up in Michigan and graduated from the University of Michigan in 1938 with a degree in mechanical engineering. His first job was with Briggs Mfg. Co. in Detroit, and from there he enrolled in Harley Earl's short-lived school for stylists. After a few months in one of GM Styling's special design studios, he was assigned to the Buick studio, where he worked until the beginning of World War II. After the war, he re-entered GM via Fisher Body, then returned to Buick, where he helped style the 1947-50 Buicks. MacKichan later joined the Chevrolet studio under Ed Glowacke and, on 2 April 1951, when Glowacke was reassigned as Cadillac studio chief, Mac became head of the Chevrolet studio. In 1962 he was named the first American director of design for Adam Opel AG in Germany. Upon his return to the U.S. in 1967, MacKichan became GM executive in charge of international design and advanced engineering. He retired in 1978.

Interview date: 21 Jan. 1975.

Lamm: Was the 1955 Chevrolet a copy of the Cadillac? Also, was it a difficult car to design?

MacKichan: There were a lot of early designs that were thrown out; there always are. Some of the early designs were quite exotic, and we always go through that stage.

I think one of the strong points in the development of the 1955 car was that it was all new. It was completely new. In terms of exterior and interior design, there wasn't anything carried over. It was a complete change from what was an old man's kind of vehicle—the 1954 and back—and it suddenly became a young man's car, with the innovative new V8 engine...that's one of the best V8s ever done. And we tried to carry that flavor, the flavor of youth, into the design.

L: Was that a joint effort between Design and Engineering?

M: A great deal of the credit should go to Ed Cole. We worked very closely with Ed Cole on that car. This was his first opportunity to get in and help on a total vehicle.

L: Of course, his background *was* Cadillac. That's why I'm wondering....

M: I don't know whether there was any connection [between Cadillac and Chevrolet], but if you look back over the years, you'll see that Chevrolet almost always has had some Cadillac influence. Now that might even go back further than my time, when Ed Glowacke was there. He was an ex-Cadillac man, and he might have purposely tried to get a flavor.... I know it's happened since then.

L: I can think of the 1932 Chevrolet, which looked like a miniature Cadillac, with all the same cues, and the 1939 had the same prow front end....

M: I don't think "copy" is the right word [for the '55]....

L: No, but it has that Cadillac look; something about it.

High fenders, low hood of early styling approach used kicked-down beltline behind windshield pillar. Kickdown was echoed in sidespear.

M: You know, people in the studio would say, This is a small Cadillac. They'd have that in mind when they'd work on the car.

L: I was wondering, too, about the '55 Chevy's dipped beltline. Where did that come from?

M: I really can't give you anything on that, because that sort of thing was just so prevalent around here. This was the sort of thing that [Styling vice president] Harley Earl really promoted. He didn't like high rear quarters on the roof. So in an effort to reduce the height of the backlight without having too high a belt, he resorted to this sort of a scheme, where the belt would hop up over the rear wheel.

L: Was his influence strongly felt on the 1955 Chevy?

M: Oh, yes. As a matter of fact, on the front end, that was almost purely [the result of] Harley Earl's wishes. And what he was thinking of was a Ferrari. He used to go to Europe every summer and spend a couple of three months over there, and he was aware of European design long before almost anybody else in this country.

L: Did you ever go with him?

M: No, I didn't, nor did other people from his staff. He used to go by himself. He came back and told us just how to execute that grille. He went into the thickness of the blades and how many blades. We made a number of different designs before we got it just right, but he was basically responsible.

L: Do you feel the grille integrated into that design all right?

M: I do, yes. He [Harley Earl] worked very closely with us on the whole thing. But that car was quite refreshing.

Clare MacKichan

Interview date: 24 Jan.1975, second conversation.

MacKichan: To get back to the car, I talked to four of the guys who were with me in the studio at that time. We met and talked about the Cadillac influence at Chevrolet, and one of the fellows recalled that Harley Earl had made the statement to Harlow Curtice in the design auditorium, looking at the 1955 Chevrolet car...he said, Now *there*...there's a car that, if it had a Cadillac emblem [on it], I could sell as a Cadillac.

There was that feeling throughout the car. You know those headlights...the brows or hoods over the lights? Well, that was something that had been on Cadillacs prior to 1955.

The eggcrate grille, as I mentioned, was really Ferrari. We were working on front ends in the studio in the early stages, and one day Harley came in and said, You know, I saw a Ferrari, and I'd like to try that kind of grille on the car. And then he described the one he wanted and, of course, we got out magazines and pictures and started to do this grille. There was a conversation about, should the sides tip out a little bit, rather than just being vertical...tip out at the top, tip out at the bottom? There were sketches made each way, and we finally chose the one on the car, which tipped out a little bit at the top.

GM's financial staff favored a divided wraparound windshield to lower glass replacement cost. Cole, Earl, MacKichan opposed this idea.

L: Was there a temptation to put fins on the car?

M: No, I don't think so; not on this car. But the tail lights were designed to give a Cadillac look from the rear. They're a little fatter than Cadillac's were at the time, but they were into the fenders in the same general fashion so it wouldn't look too Cadillac-ish, except to put that angle on the sides. On the interior, you know, the whole instrument panel was fashioned after the Corvette. So we just derived that from the Corvette was best we could. You remember the bowties across the insert in the panel? There was a great deal of work on that, getting the bowties precisely right. We really were big on bowties in those days, had more bowties than Pontiac had Indians. On the hubcaps and wheelcovers, we had a whole circle of bowties.

Early in the program, this car was being worked on down in Chuck Stebbins' body room. Stebbins' designers had a typical, conventional car going, with a very heavy roof; even heavier than the 1955 Chevrolet came out with, and the '55 was generally a fat car. Stebbins' model was shown to Cole...Harley Earl showed it to Cole...and Earl had wanted to get the car lower than normal; lower than the 1954 Chevrolet by quite a bit. Earl said, I'd like to get that s.o.b. down to 61 inches, which sounds high today. So they made a clay model that was 61 inches high, and Earl said, I can't wait...I've got to show that to Ed, and he brought Cole in, and he said, Here, we've got this car for you that's only 61 inches high...*61 inches*! And Cole said, Hell, I wouldn't want to make it over 60! That sort of typified the feeling in those days: each person wanted to out-do the other.

But boy! This was really Cole's car, and he spent a lot of time in the studio...took a very personal interest in the whole thing.

L: Tell me something about the body development room.

M: Generally that's where they'd lay out the basics of a car: the wheelbase, width, length, etc. We in the studio could change the overhangs, the height, and the sculpting.

L: Did the body room get involved before the studio actually started on the car?

M: Not necessarily. They might start a little ahead of us, but we would be working on the exterior sheetmetal and, to some extent, on the interior.

L: And would the studio keep working with the body development room as the project progressed?

M: That's right.

L: And how were the studio and the body room connected physically? Were they next door to each other?

M: No, the body room was on the floor below us.

L: Let's talk about the executive coupe. How far did you get with that? How seriously was it considered for production?

M: It was in the studio, but I don't think we ever engineered it.

L: The executive coupe had those two inward-facing seats in the rear. Were those jumpseats that could be folded into the side walls?

M: As I recall they were. But it was basically a 2-passenger car.

L: Was there some way to reach the rear-seat area from the trunk?

M: I don't think so. I don't really remember.

L: Were there any other unusual body styles for '55?

M: Well, the Nomad. Those two were the ones.

L: What else comes to mind about that design?

M: Cole wanted the whole car to typify the same sort of feeling: youth, speed, lightness.

L: Who were some of the other people involved?

M: On body engineering, there was Jim Premo. Carl Jacobs worked for him....

L: And who were some of the other people in your studio?

M: Carl Renner, of course, [who did the Nomad] and Duane [Sparky] Bohnstedt [designer], Ed Donaldson [soft trim and colors]. Chuck Stebbins had charge of the body room. Then there was Hans Hierta, who's now with Chrysler. Actually there's a picture that shows all of us; everyone in the studio.

Clare MacKichan

Interview date: 15 Jan. 1991, third conversation.

Lamm: Now on the '55 program, some of the pictures date back to a time before Ed Cole got to Chevrolet. Cole arrived in May 1952, and some of

these pictures, which look to be genetically linked to the production car, are from before that. You can see some of the embryonic '55 coming out, yet it was quite a different car at the very beginning. For instance, the hood looked like it stood lower than the fenders, and the fenderline dipped down at the windshield pillar instead of on the rear beltline.

MacKichan: I think we went through that phase. The way things were done then at General Motors Styling, the initial work was done in the body room.

L: Some of the photos simply show dimensions. They're of cars in plan view and side view. One's a comparison with a 1952 Ford. It's the 1952 Ford and an overlay of the 1955 A-2, it's called. I guess that's the A body, and I don't know what the 2 means. Would those have come from the body room?

M: Yes, that would be a normal way to do things, comparing our cars dimensionally with other cars.

L: And would those comparisons have originated in the body room?

M: Very possibly.

L: Was the Ferrari grille something that Mr. Earl introduced fairly late in the 1955 Chevrolet program?

M: I don't think it was late. It came about in the middle. But, you know, we were encouraged to try all sorts of things. Flights of fancy.

L: And I understand that the '55 didn't sell that well in the beginning and that you were asked to do a crash program on an alternative grille, because the sluggish initial sales were blamed in part on the Ferrari grille. But I understand that by the time you got another grille drawn up, sales had taken off again, and there was no need to put a different 1955 grille into production. Do you recall that?

M: Vaguely. Of course, we were working on the '56 by that time.

L: I understand that the crash-program 1955 grille became the '56 full-width front end. But let's stick with the '55. On the 1955, apparently the interior was done in your exterior studio. I believe you guys had responsibility for the interior as well as the exterior.

M: Yes, except for the soft trim. The soft trim was done in Ed Donaldson's area.

L: What about the 1955 Nomad?

Motorama 1954 Waldorf Nomad inspired so much favorable public reaction that Chevrolet introduced first production Nomad in Jan. 1955.

M: I can remember very distinctly. It was a Saturday. We were working one Saturday, as we usually did. This was during the time of the Waldorf show in New York. Howard O'Leary [Harley Earl's assistant] called me up, and he told me that they just made the decision to turn the Nomad into a production passenger car, not based on the Corvette but based on the 1955 car line. O'Leary said, You guys get busy and show us on Monday when we come in what it would have to look like.

So we immediately got out the Corvette Nomad drawings and cut them up and added features here and there...moved things around, so by Monday we had a good drawing of the production Nomad. It had the fluted roof and the bananas on the tailgate, the wraparound rear glass....

L: One thing I've always marveled at, and something that doesn't get much attention, are the Nomad's full rear-wheel cutouts. You didn't see that on station wagons before the Nomad came along.

M: The production Nomad had them because the Corvette Nomad was that way, and we wanted to keep as much of the original car as we could. We made the production Nomad as close as we could to the Corvette Nomad in every aspect.

L: And that wraparound glass in the back? It must have been expensive.

M: It was, but it appealed to Earl and Cole. Cole was a great one for not paying any attention to costs until he got what he wanted, and then he would try to get the cost down. He fought endless battles with the cost people. And if he didn't convince them the first time, he'd try again. Usually the weight of his arguments and his stature in the corporation would force them to do it his way. But he was never one to worry about costs until he got what he wanted. That was one of the reasons—one—that Earl admired him so much.

L: What were some of the other reasons?

M: Well, Earl admired Cole for the fact that he was a very progressive product kind of guy. And he had pretty good taste.

L: And he knew what he wanted.

M: If he didn't know what he wanted, he would wait until you produced something he did want. And if he didn't like something, he was pretty nice about it. He wouldn't get all excited and make a big fuss.

L: Unlike Mr. Earl.

M: They were on a little different level. Nobody called him Harley. It was always Misterl. Cole was just Ed. Even in the studio. He didn't want to be called Mr. Cole. And he would come into the studio and joke with everybody—the modelers and the engineers. He was a wonderful guy. He had a wonderful way with people. Everyone admired him and wanted to please him.

L: And he understood the processes, too, I imagine.

M: That's right, and if someone didn't know how to do something, he'd tell them. Ed had a lot of enthusiasm, and that would transfer to anybody he was talking to. ❑

Designer Carl Renner successfully transferred Waldorf Nomad's fluted roof, wraparound rear glass, wheel cutouts, and tailgate "bananas" to the 1955 passenger car, as shown in early Fisher Body prototype.

Renner also rendered this and the other two sketches on these pages. They were all made after abandonment of the low-hood theme.

Dipped beltline began to appear here. Pontiac borrowed air intakes above headlights, while Olds took possession of rear fender sweep.

The 1957 grille begins to take shape in this 1953 sketch. Both 1956 and 1957 front ends were done at the same time as the 1955 grille.

Carl Renner

Assistant chief designer, 1955 Chevrolet.

Carl Renner was born in Germany in 1923. His family came to Detroit when Carl was four. At age 20, he landed a $25-a-week job as a cartoonist for the Disney studios in Hollywood. He'd always wanted to design cars, so right after the war, he took himself back to Detroit and found work with General Motors, first in a special advanced studio and then, in 1946, in the Chevrolet studio, first under Ed Anderson and later under Ed Glowacke.

Renner became Clare MacKichan's Chevy studio assistant in 1951 and took an active part in the 1955 model's entire design process. His contributions to the '55 included the Nomad wagon, the dipped beltline, the eagle hood ornament, and the deluxe steering wheel. Renner later worked in the Olds, Pontiac, and advanced studios, where he played major roles in designing the Motorama LaSalle II roadster and the Cadillac Cyclone. Carl retired from GM in 1980 and again lives in Southern California.

Interview date: 30 Aug. 1975.

Lamm: Tell me how the production Nomad got started. Was it with the Corvette Nomad?

Renner: The Corvette Nomad station wagon was requested by Harley Earl, and at that time I made a proposal sketch for it. Mr. Earl requested a Corvette fastback for the 1954 Motorama show, and he thought it would be nice to have an unusual station wagon, so I was with the Chevrolet studio at that time as a designer. All of the designers made proposal sketches, and Mr. Earl happened to pick the one I did, which ended up, as you saw, as this Corvette station wagon. And that's how it came about. From there we jumped to the 1955 Chevrolet.

L: So it went directly from the Corvette Nomad to the production 1955 Chevrolet Nomad?

R: Mr. Earl decided that the 1955 line should have something special and different, and he came into the Chevrolet studio at the time. Yes, he said he wanted the Corvette station wagon upper put on the 1955 body. This was in 1953.

So working on that specific design, I got the job under Clare MacKichan, who at that time was the chief designer of the Chevrolet studio. And with his direction and Mr. Earl's, I took the print of the Corvette station wagon upper and spliced it—cut it apart and adapted it to the lower of a 1955 Chevrolet, using that windshield and extending it and so on.

L: How did the roof fluting come about?

R: The original thought was to have the forward portion of the roof and the second body pillar painted in body color. Mr. Earl loved brushed aluminum and stainless steel, and the idea was to have the back of the wagon roof

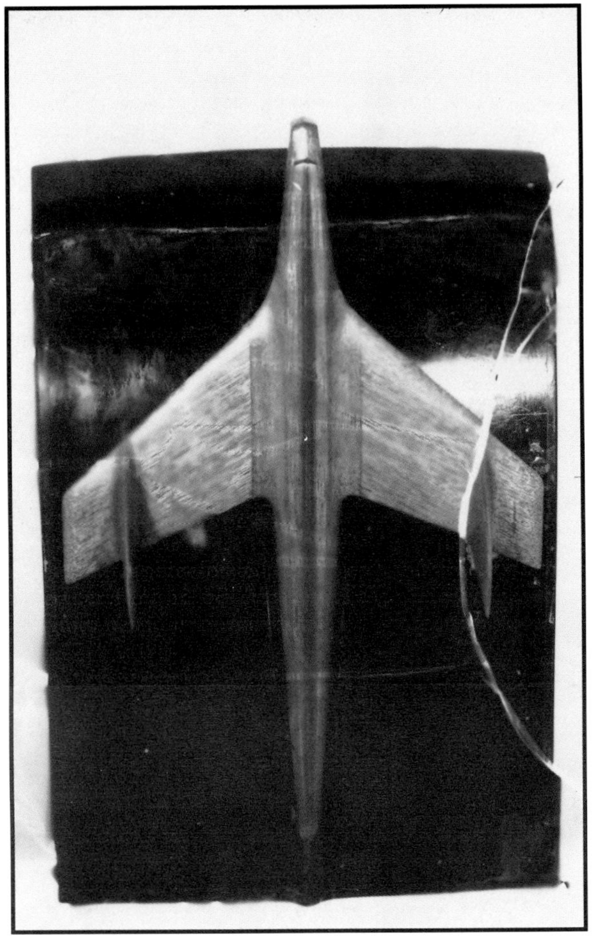

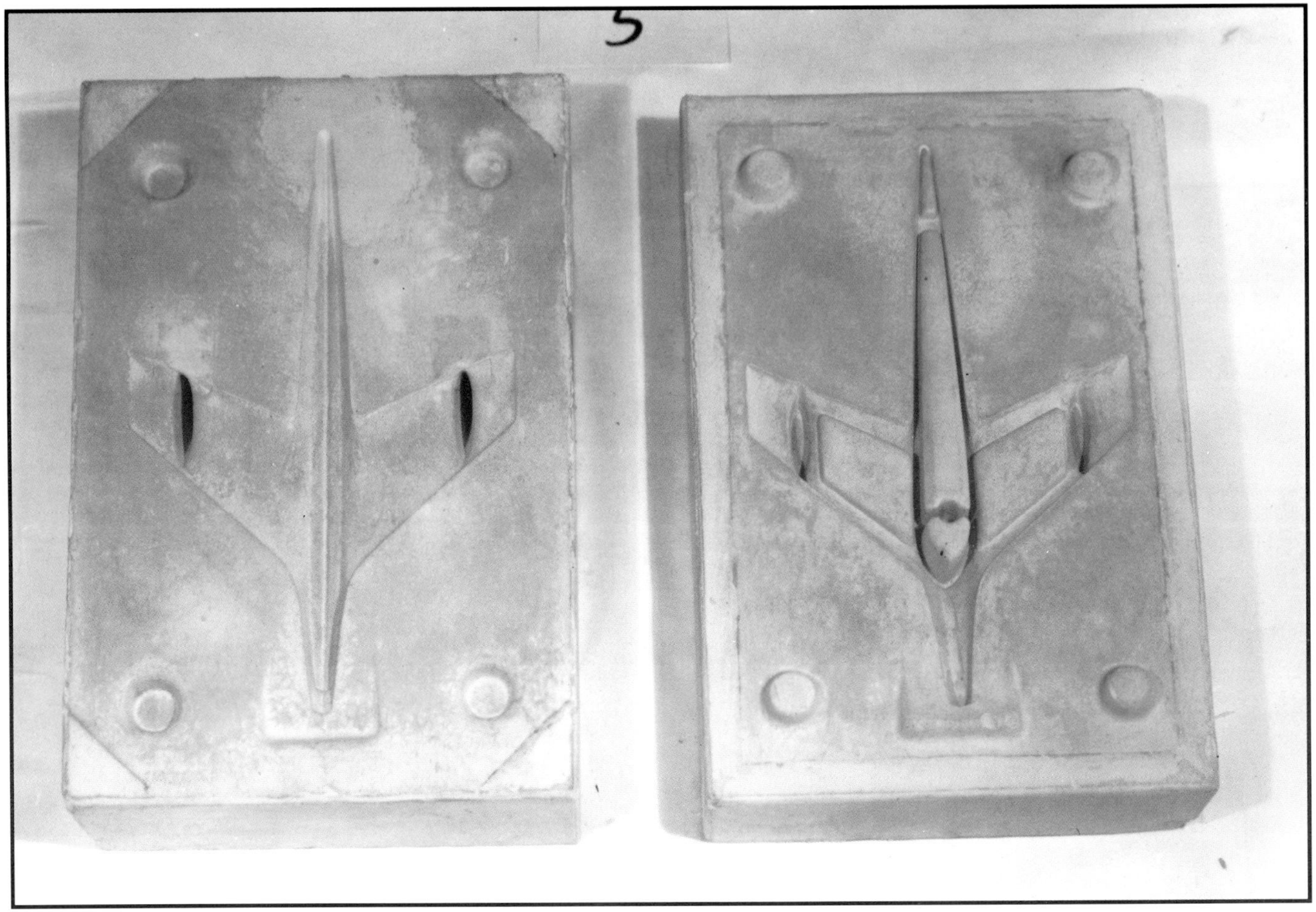

Renner did all the early-50s hood ornaments, shown here in the form of a mahogany die model and preliminary casting molds.

made of unpainted metal and to slide it open or retract it like one of those little metal drinking cups that collapse in on themselves. So that's how those lines across the top came about. The retractable feature was dropped, but the scribe lines stayed.

L: Was there anyone else involved in the production Nomad design?

R: Jack Park at that time was the Chevrolet studio chief modeler. Chuck Stebbins had played with those rub strips on the tailgate. We called them bumpers or bananas. Many guys played around with decorating the trunklid.

Carl Renner

Interview date: 12 Jan. 1991, second conversation.

Lamm: I understand that, in addition to the Nomad, you also had something to do with the 1957 Chevrolet front end. Is that correct?

Renner: Yes. I had the 1957 bumper-grille. The grille for the 1957 was done in 1953. Everyone was going for bumper grilles back then: Buick, Olds. They happened to like that design [for Chevrolet], and it did eventually go into production, but it was very costly.

L: The '55 grille must have been expensive, too.

R: Not as expensive as that bumper-grille. It was very unusual for Chevrolet to go to such a massive, husky-looking grille.

L: Why do you suppose I can't find any sketches or clays showing the 1955 grille's development?

R: Probably because there weren't any drawings or sketches made. Mr. Earl said he wanted a rectangle in there, and it went into clay, and he said, Now we want a bunch of bars in this way and that way, the eggcrate pattern, and he specified how far apart, and that was it. There were no drawings or renderings.

United States Patent Office Des. 181,593

Patented Dec. 3, 1957

181,593

AUTOMOBILE

Clare M. MacKichan, Farmington, and Joseph R. Schemansky, Birminghâm, Mich., assignors to General Motors Corporation, Detroit, Mich., a corporation of Delaware

Application August 25, 1954, Serial No. 32,016

Term of patent 7 years

(Cl. D14—6)

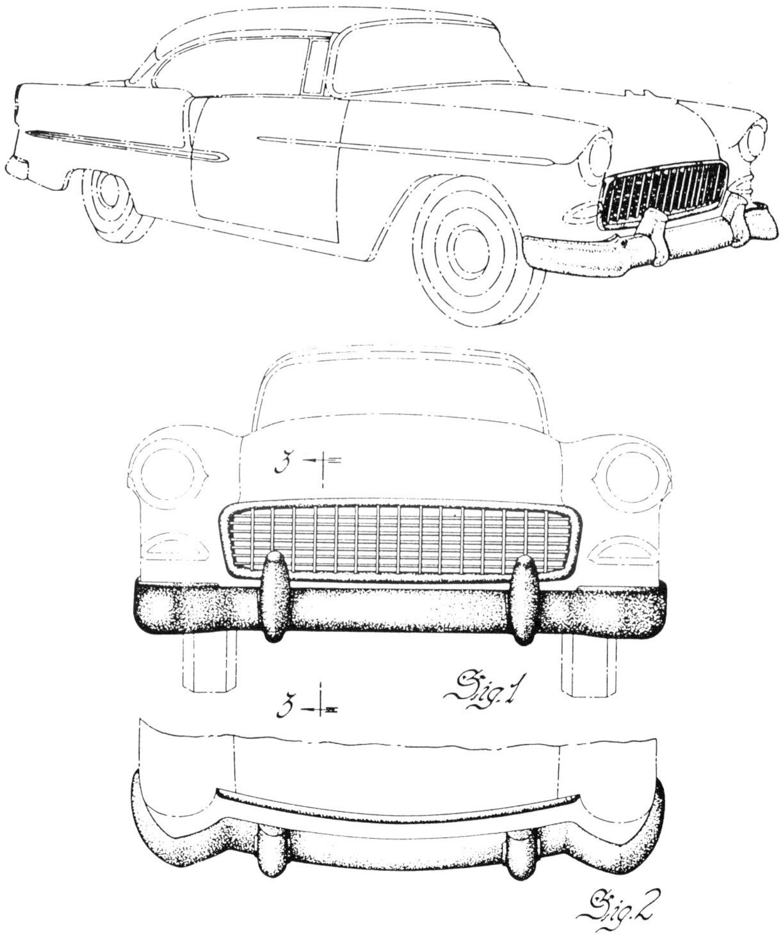

Harley Earl suggested eggcrate grille after viewing Ferraris in Europe. MacKichan, Schemansky applied for design patents in 8/54. Grille was later accused of slowing sales startup.

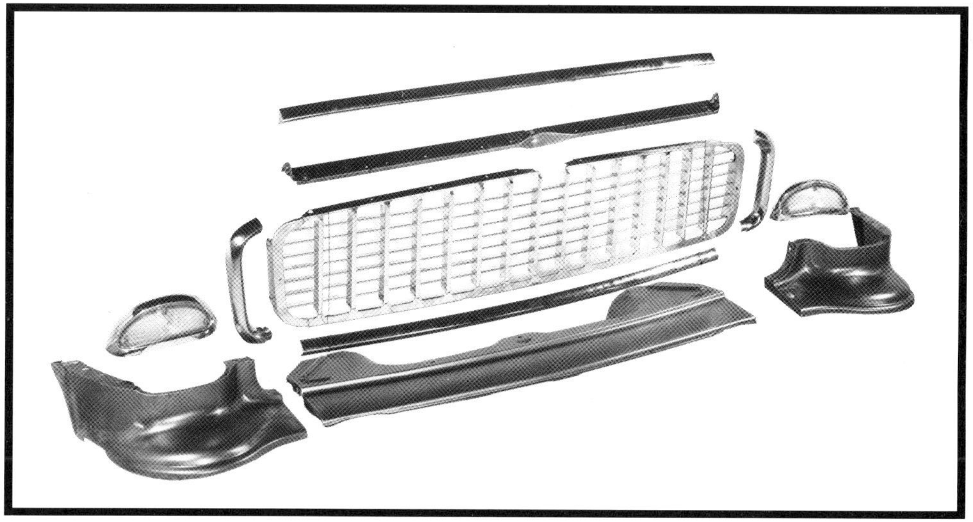

L: Was that fairly late in the program?

R: Not too late, but we tried many, many grilles in there at first. Mr. Earl wasn't satisfied, as I recall. But the reason it came in so quickly was because, before the raised hood we had a dropped hood. And I don't know how many grilles we did on that. So that took a lot of time before we went into the final '55.

But, Mike, I want to point out one important thing. Although one person might create a certain design, the total outcome is due to the talents and efforts of many. You have to understand that lots of other people contributed to these cars. It's not just one man alone. Mr. Earl was the leader; he made the decisions. We have to give credit where it's due.

What a lot of us did was called CDS: created, designed, and supervised. For example, the Nomad upper. Mr. Earl wanted something new in the form of a station wagon. I created it, I designed it, but I didn't supervise it. But I also have to give credit in the design to the tech stylist; the line man. He sweetens up the lines for you. You just rough it out.

Then the modelers have to get credit also. They enter into it and try to save the design when it goes into three dimensions. Sometimes one angle, the side angle, or the front section angle, will have to be changed, because it's like a mitre with a tube in it. Those surfaces come together, and maybe they become distorted, so the modelers have to work it out in 3-D. A modeler can make or break something important.

L: I understand you did parts of the instrument panel and the steering wheel and the column.

R: I remember doing the steering wheel.

L: Tell me about the body development room.

R: That was under Chuck Stebbins. Chuck Stebbins had a lot to do with the 1955's body development. He also created the 1955 tail lamp.

I have to tell you, though, about the dropped hood. In the very early sketches, the body from the cowl forward was a high-fendered style with a dropped hood. The hood stood lower than the fender crowns, like the Nash. That theme was going ahead strong for a long time. The front fenders went straight back from the headlights to the windshield pillar. At that point, the line dropped down. In other words, the beltline stood lower than the front fenders.

Anyhow, it was in clay, and we stayed with that theme for a long time, but it just didn't look right. It didn't work out. The hood was low, and it looked squashed. Mr. Earl went back to a nice, big, substantial hood.

L: Now as I understand it, you did the 1955 accessory bumper guards.

R: Yeah, on the '54s, too.

L: Who did the Nomad eyebrows?

R: Bob Veryser.

L: And who did the Nomad's full rear wheel cutout?

R: That would have been done in the body room under Chuck Stebbins. By the way, I'd like to mention that the hood and decklid emblems for the 1955 and '54s...those were done by Bob Bartholemew. He was Chevrolet's crackjack man with details and emblems. You might want to mention that. Bob Bartholemew.

Before I forget, the side-treatment slash on the rear fenders of the '55 was done by Bill Steere. The dipped beltline was done by me. General Motors patented that; a design patent. They also patented the steering wheel design. ❑

Edward J. Donaldson

1955 Chevrolet interior designer.

Ed Donaldson was born in Kingston, Pa., in 1933. He attended Meisinger School of Art and Wayne State University, both in Detroit and, in 1954, returning from the Korean War, Donaldson joined General Motors Design Staff as a designer and illustrator. He became chief designer of Chevrolet interiors in 1956. Ed's career has included assignments with GM's British subsidiary, Vauxhall; he's overseen various showcar designs, and he's managed advanced studios, environmental systems, and production interiors.

Interview date: 21 Jan. 1975.

Lamm: I was wondering, first of all, whether you worked on the first Corvette, the 1953 model?

Donaldson: No, I didn't.

L: I ask because it seems to me that the 1955 passenger-car instrument panel design borrows from the first Corvette.

D: Right. They were very similar.

L: Was that Corvette theme carried over on purpose?

D: In a way, yes, it was. In fact, the instrument panel was done in the exterior studio at that time. The interior studio was doing strictly soft trim.

L: Oh? Who was in charge of interiors in the exterior studio?

D: Well, Mac [MacKichan] was in charge of the entire studio. I'm not sure who did the instrument panel. It might have been Joe Schemansky. [Pause.] I just talked to Sparky Bohnstedt, who was also in the studio at that time, and he says that Carl Renner did most of the work on the 1955 instrument panel.

L: So your responsibility was mostly soft trim?

D: Right, and I also worked on the Nomad soft trim.

L: That was the waffle upholstery pattern?

D: Right. Actually that was practically the beginning of the dielectric [embossing] process. Prior to that there was a lot of saddle stitching and novelty things like that. But I think the Nomad was the first breakthrough on a real dielectric process, and at that time it was pretty darned good.

L: Apparently they don't use dielectric much anymore, at least not in the same way.

D: Well, we've gone through a cycle. I remember I used to be really gung-ho for [dielectric embossing], because I thought we could do a lot of things that couldn't be done on a sewing machine. And we did have some problems with Fisher Body management, because they had to change over their plants to facilitate the dielectric machines. It was a new thing, of course, and you had to develop systems and reliability testing. We got a couple of guys over there who really knew what they were doing, and they did a fine job.

But then after a while it reversed itself, and we went back to cut and sew. It looked a little more finished and custom. If you look at interiors today [1975], whether it's our product or Ford's or Chrysler's, everybody's getting really posh. They're doing it up so you're getting more than just the old standard, ordinary bench seats with sewn pipes on them.

L: I understand there was originally an executive coupe planned for the 1955 car line. Do you remember it?

D: I remember the name. [Aside.] Sparky, do you remember anything about the 1955 executive coupe? [Returns.] You know, we've done so many blasted cars....

L: I know. Well, that was the coupe with two inward-facing rear seats. I think they folded down. I was just wondering whether they were anything like bucket seats.

D: I honestly don't remember. I recall the name, and I think that was almost the code name for the first Impala, too.

L: Really?

Base Del Ray instrument panel had plain inset, no stamped-out bowties. Double bubble gauges came from Vette.

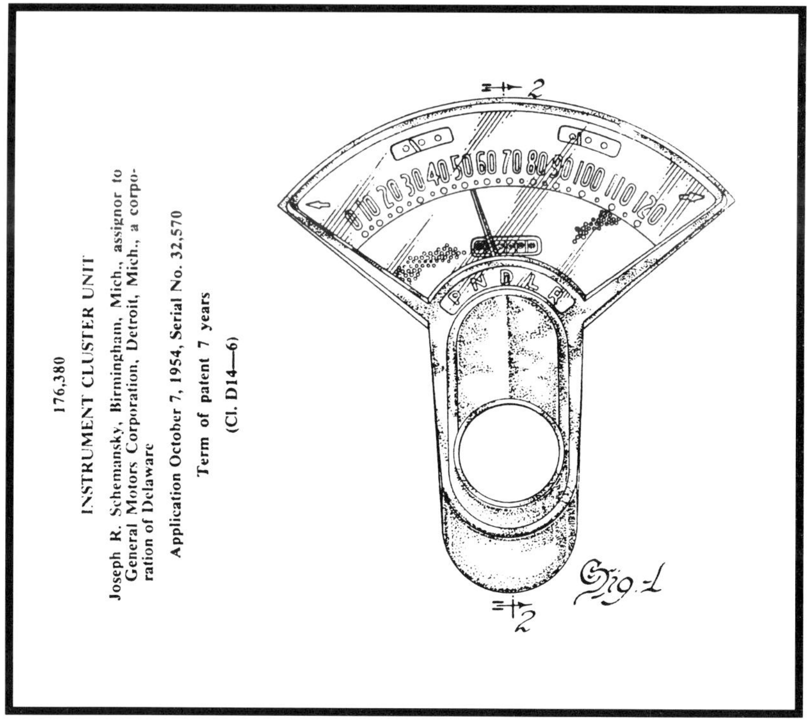

At Ed Donaldson's behest, production Nomad interior used dielectric embossing on seats and side panels. Chromed top bows repeated outer roof grooves.

Design patents mean little but do establish priority. Assistant chief Schemansky made patent applications.

D: We did an Impala Motorama showcar for 1956 or '57. When we developed the 1958 Impala, it was called the executive coupe for a time. This was a low coupe; lower than normal. Those were real exciting years, and we had a lot of fun doing those cars. I wish I had a 1955 right now!

Ed Donaldson

Interview date: 23 Jan. 1991, second conversation.

Lamm: You guys in soft trim must have had a separate studio somewhere; separate from MacKichan's exterior studio.

Donaldson: That's right. We were on a different floor down in the old building. In those days, the exterior guys did the instrument panel. We interior guys did the soft trim and the exterior and interior colors. Because, you know, the interior and the exterior have to be coordinated. It's still that way. That responsibility fell to our color and trim studio. At that time Henry Lauve was in charge of all interior studios. Then each of the division interior chief designers were his assistants. And I had Chevrolet at that time. ❑

Duane L. Bohnstedt

Designer in Chevrolet studio.

Bohnstedt's dad was a Chevrolet service manager in Beatrice, Nebr., and that's where Sparky grew up. He attended the University of Colorado at Boulder and later went to Art Center College of Design in Los Angeles, graduating in late 1951. He joined GM Styling that year, in the Oldsmobile studio. In the fall of 1952, he transferred to the Chevrolet studio. Bohnstedt left GM briefly to work with Dick Teague at Packard, where he was chief designer of the Clipper studio. He returned to GM, became assistant chief designer at Pontiac, taught at Stanford University in 1961, joined Clare MacKichan at Opel in Germany in 1963, and stayed overseas for a while with Mack trucks. On his return to the U.S., Sparky became GM's vehicles coordinator for the 1964 World's Fair. His last position was as chief exterior designer for commercial vehicles. Bohnstedt retired in May 1984.

Interview date: 28 Jan. 1991.

Lamm: Let's talk specifically about the 1955 design.

Bohnstedt: My first job when I went into Mac's [MacKichan's] studio was to lay out the instrument-panel insert with all those little teeny bowties. I made an etching drawing about seven or eight times actual size.

L: How many bowties did you have to put on the master?

B: I couldn't tell you. I made the thing up about 10 inches long by whatever the proportions were. I laid it all out and did the drawing in India ink. Then I reduced that to the right height. The original wasn't debossed and filled in with black; it was actually punched out. I still have some of those little Chevrolet bowties punched out of aluminum in a box around here. The box was in my desk at work.

I think the guy who was really responsible for the instrument panel, though, was Carl Renner. The other fellow in there at that time was Bill Steere. I know that Carl did the steering wheel and some of those things. I think he did most of the interior. We did the panel in the exterior studio at that time, but not the seats or the colors. That's when we were still back in the B Building, on the 10th floor, I believe.

L: How did the studio and the body development room work together? Was Earl the only liaison?

B: I think the body [of the 1955 Chevy] basically would have been done in the body room. We'd take the basic body in the studio, and we did all the front-end sheetmetal, the rear end, the lights and all, but the basic body was Chuck Stebbins'.

L: Tell me about some of the details.

B: When it came to accessories, there was a fellow by the name of Bill Beuchler. We called him Bubbles. He did all the accessories. We "pure" designers didn't want anything to do with those. He did an aftermarket hood ornament with a plastic bubble over it. That's where he got the name. Bill did all the rocker moldings and the continental kits, the bumper ends, those kinds of things that weren't very pure. Most of us didn't want anything to do with them. Bill Beuchler did those. ❑

Bob Cadaret

Designer, 1955 Chevrolet studio.

Born and raised in Los Angeles, Bob Cadaret began drawing cars at age five. At 18, he won the coveted Fisher Body Craftsman's Guild award and a $5000 GM scholarship. He graduated from Art Center College of Design and entered GM Styling in 1953, coming in on the end of the 1955 Chevrolet program. In 1956, he came up with the names Monza and Lakewood. His more major contributions had to do with the 1957 Chevrolets, the 1958 Impala, and the Nova. After 25 years with GM, he retired in 1987 and now devotes much of his time to painting.

Interview date: 28 Jan. 1991.

Lamm: I understand you had something to do with the Waldorf Corvette Nomad.

Preliminary braces and pedal suspension began to take form in Apr. 1953 on wood and metal mockup. A 1953 steering wheel perches atop wooden steering column. As such components as heater box and other under-dash components were fitted, they were placed into this buck.

Cadaret: A little, not much. It was almost done. I got to put the little exhaust pipes on the back fenders, and I made quite a few sketches on the upper, but Carl Renner's were chosen and put on the production cars, as you know.

L: Do you know who did the original double-hump instrument panel on the 1953 Corvette? That was the theme carried over to the '55 passenger car and the trucks. Who did those?

C: Bob Bartholemew, I believe. Bartholemew did that panel on the original Corvette, and then it just worked its way into the passenger cars. Bob later became chief of the Chevy interior studio.

L: When did you get to the Chevrolet studio?

C: Around June of 1953. The only thing I got on the '55 was the little chrome piece on the back of the fender where the 2-tone paint breaks. As a young designer, I was very happy to have contributed even that much.

L: That was really important. I had always felt that the '55 Chevrolet took a lot from Cadillac. It was like a little Cadillac.

C: That was purposeful. Harley Earl often mentioned the similarity with Cadillac when we were doing the 1955, 1956, and 1957 Chevys. He wanted the average guy to feel he was driving a baby Cadillac. And that's why some of the things that wound up on the 1955 Chevy look like they came from Cadillac. In those days, the average guy couldn't afford to even look at a Cadillac.

Ed [Cole] was in the studio a lot. A lot of general managers after him didn't come in the studio that often. Cole and Earl were there almost every day. They worked right along, talking and planning with the designers, and we had a fantastic combination there—the two top guys plus the designers, and that's why we were able to work out the things we did. We got decisions made real quick. In later years, the design committees came into being, and you couldn't get as much done as fast.

I always thought the 1955 Chevy had very clean lines, and with its great 2-tone paint schemes plus the new V8, the car was a milestone of that period. ❑

Wooden heater-box mockups, the lower one with a side-plenum air inlet, took several forms before engineers decided on a final configuration.

Edward N. Cole

Chief engineer, 1955 Chevrolet.

Born in Marne, Michigan in 1909, Ed Cole grew up expecting to become a lawyer. An early hot-rod Saxon roadster, though, turned him toward cars and, in 1930, he received a grant from Cadillac to attend the General Motors Institute in Flint. He went to work in the Cadillac dynamometer room before his graduation and, by 1946, had worked his way up to become Cadillac's chief engineer.

There, he and his co-engineer, Harry Barr, oversaw creation of the highly influential 1949 Cadillac overhead-valve V-8. That engine helped set the scene for the 1955 Chevrolet V-8. In 1950, Cole moved to GM's Cleveland tank plant for the duration of the Korean War, then became chief engineer of Chevrolet in 1952, where he and Barr oversaw development of the 1955 passenger-car and truck lines, including the small-block V8.

By 1956, Cole had become Chevy's general manager. He was made a GM group vice president in 1961, executive VP in 1965, and president of General Motors in 1967. He retired at 65 (GM's mandatory retirement age) in 1974. After retirement, he became involved in an international airfreight venture and also with a modernized, redesigned Checker cab. Ed Cole's private plane crashed during a storm in May 1977, and he was killed. This interview was originally published in Special-Interest Autos *magazine in March 1975. Reprinted with permission.*

Interview date: 13 Jan. 1975.

Lamm: Mr. Cole, was there a V-8 in the works at Chevrolet before you got there in May of 1952?

Cole: Yes. Chevrolet wanted to upgrade their product. And a V8 was underway by the then-current engineering organization. They had an engine in the works at that time. When I was asked to come in, I was also asked to take a look...at everything that was going on for the new car.

Everybody recognized the need for a V8 and a new car, and I didn't feel that [what they had at that point] included some of the latest technology as far as the manufacturing techniques or its design charactertistics were concerned. In other words, I felt there were other ways to make that engine; make it better; more precise and less costly.

L: What did they have on the drawingboards?

C: Well, they had an engine similar to the 1949 Cadillac V8. It was a rather heavy engine. And costly [to manufacture]. That was when we decided to make the precision cylinder block, which became the heart of the engine. [We did this] by using an entirely different casting technique. We used the green-sand core for the valley between the bores—the 90° vee angle in the center. We used a green-sand core to eliminate [most of] the dry-sand cores [used previously]. Getting rid of the dry-sand cores let us cast the block upside-down. That allowed the plate that held the bore cores to be accurately located, so the wall thicknesses could be controlled. And we could cast down to 5/32nd-inch jacket walls. That was the technique [we used].

New front frame rails and crossmember were also previewed in wood. This version included angled braces that weren't needed in production.

Once Styling released the 1955 body design, each individual panel was fabricated in fiberglass. Wooden jig retained checking templates.

Fiberglass outer skins were then mated to cowl, floor, and other panels. Metal prototype parts were installed as they became available.

And then we went to the stamped rocker arms and the hollow pushrods for a different oiling system.

L: Was the green-sand method your idea?

C: It was a combination of ideas. I think John Dolza, who was with GM Engineering Staff, had as much to do with that as anybody. But none of these things are individual ideas. They come out of brainstorming sessions. One idea feeds the other.

L: What about the hollow pushrods?

C: Those had been used before, but it was slipped into the picture along with the stamped rocker arms. [The idea was to use a] self-aligning rocker arm with respect to the valve stem to reduce the thrust on the valve as much as possible—the thrust between the valvestem and the guide. It's now being used by several manufacturers, but it was an innovation we incorporated in that new V8 for 1955.

L: Your predecessor as chief engineer of Chevrolet was Ed Kelley, wasn't it?

C: Yeah, Ed Kelley.

L: So the former V8 that they'd been working on was done under his direction?

C: That's right.

L: Were there any of his staff people who came over and worked on the V8 as you conceived it?

C: They all did. I just brought in Harry Barr and two or three others who had the burning desire to get the job done. What happened...Chevrolet had been working on a V8 for some time. And when you do that, you lay down a design and you start making just little changes on an existing design without starting with a clean sheet of paper. What we did was to start with a clean-sheet approach. We did that with the entire car.

L: Well, it's interesting that they were copying the 1949 Cadillac V8, because that was your baby. So you had to take it one step beyond. Why you, though? If you conceived that original engine, why did they choose you to conceive one better than that?

C: Well, that's a long story. You see, they wanted me to come into Chevrolet as manufacturing manager. I didn't feel [Chevrolet Div.] was exciting enough that I wanted to be manufacturing manager. That's how this thing started. [Cole let it be known that he would come in only as chief engineer. Ed Kelley, who'd been Chevy's chief engineer, agreed to become the division's manufacturing manager, giving his engineering responsibilities to Cole.]

The whole concept of the car was built around lighter components. We started, in other words, to get away from the heavier torque-tube drive and went to the Hotchkiss type. We went to a Salisbury-type axle instead of the banjo type. And then we went to a balljoint front suspension for weight saving. And to a tubular frame.

Then, too, we had a brand-new body. Today, if you wanted to take the risk with Chevrolet, the guy who thought that up would probably be fired,

These are early Fisher Body sheetmetal stampings of floorboard and cowl, with the access hole showing an actual engine bellhousing.

Prototype 1955 frame, delivered in Feb. 1954, was fully boxed, ended up being 50% lighter and 18% more rigid than the 1954 had been.

Front brakes were among the few major components carried over from 1954. Nothing interchanged, though, in front suspension.

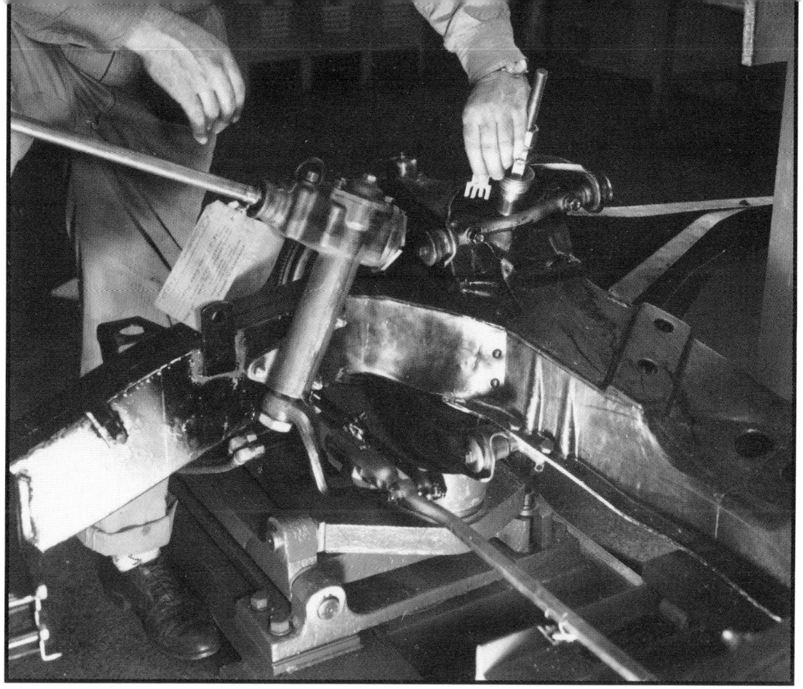

Same steering box was used with non-assisted and power steering, although power steering had a longer pitman arm for quicker ratio.

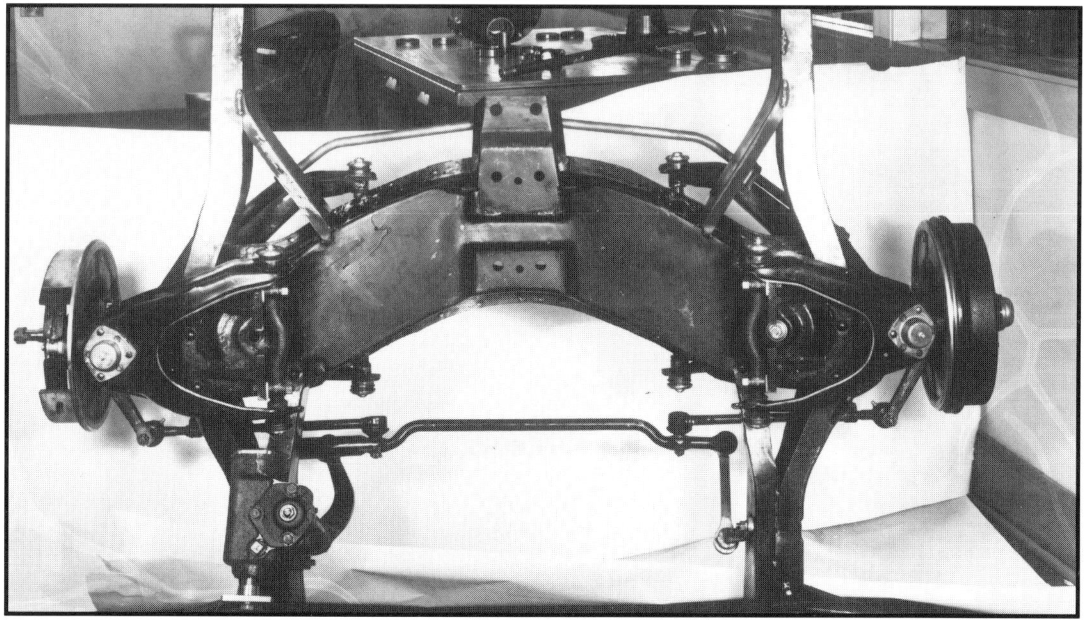

Balljoint front suspension was new to Chevrolet for 1955, and tread was increased 1.3 inches. Stabilizer bar stayed throughout new model's development but didn't appear on most models.

By Feb. 1954, Fisher Body could supply some hand-finished exterior sheetmetal stampings.

because that was a brand-new car from the ground up, and done in a very short period of time.

L: Would you consider Ed Kelley a very conservative sort of engineer?

C: No, he was.... Ed Kelley's background was heavily oriented toward manufacturing. Anybody who's been in manufacturing has a tendency to design to what the manufacturing people feel they need. [And an engineer with that background wouldn't] stretch them out perhaps as much as they could be. For example, when we talked about 5/32nd crankcase walls, and the casting for the new block, the manufacturing people said, Well, you just made a mistake. We don't make anything less than 5/16th. I've got to run, Mike. That's about it. ❑

Harry F. Barr

Assistant chief engineer, 1955 Chevrolet.

Born in Enid, Oklahoma in 1904, Harry Barr studied mechanical engineering at the University of Missouri before transferring to the University of Detroit in 1926. Summers, he worked on a Studebaker assembly line and at Detroit Edison under a work/study program. He later went to Cadillac as a laboratory technician, becoming a project engineer in charge of chassis and drivetrain in 1937. He helped develop the ohv V8 that eventually powered the 1949 Cadillac, then graduated to chief engineer of GM's Cleveland tank plant during the Korean War.

He followed Ed Cole to Chevrolet in 1952, becoming Cole's assistant chief engineer in charge of engine and chassis. Barr rose to chief engineer in 1956 when Cole became Chevrolet general manager. In 1963, Barr was named GM vice president in charge of corporate engineering. He retired from that position in 1969. This interview originally appeared in Special-Interest Autos *magazine in March 1975. Reprinted with permission.*

Interview date: 19 Dec. 1974.

Lamm: Mr. Barr, let's begin by your telling me something about your coming to Chevrolet. Was the specific reason for moving both you and Mr. Cole to Chevrolet to develop the new V8?

Barr: Well, it was to develop a new image in Chevrolet engineering. But yes, I guess the engine was one of the major reasons. Eddie Cole had been quite successful at Cadillac with the 1949-50 car. He had overall car, engine, and management experience.

L: When did development of the 1955 Chevy V-8 actually start?

B: In the summer of 1952. Ed went to work in May, and I got there in June. One of the pressing projects at the time was to put a few more good things on the 1953 and '54 Chevrolets that were coming. That was confined to transmission work plus the addition of air conditioning, power steering, things like that. But the main thrust was on the 1955 car, which included the new engine, an automatic-shifting Powerglide, new frame, new body, new everything.

L: I suppose the corporate decision to re-engineer and re-design the Chevrolet for 1955 had already been made by the time you and Mr. Cole got there. Is that correct?

B: No, I don't believe it worked that way. Really the corporate decision was that they [Chevrolet] needed a car that would go faster. In fact, when we got there they had a V8 in design, on the boards, a 231-cubic-inch, something like that. But it wasn't a short-stroke, big-bore engine. It did have overhead valves.

L: Who had responsibility for the previous V-8, the 231-cid version?

B: That was done under Ed Kelley as chief engineer. He was doing a good job, but when Mr. Cole came in as the new chief engineer, he felt that if we were going to spend all that money for new tooling, we ought to refine the engine so it had more potential for growth. Which, of course, ours did. That engine is now up to 400 cubic inches.

But basically, different groups were formed to design the various components of the new automobile. I was in charge of the chassis. Ed Cole had the engine and the powertrain parts. Jim Premo was on the body and sheetmetal.

L: Was the 1955 Chevy ever talked about as being a small Cadillac while it was being designed and engineered?

B: You mean within Chevrolet [chuckles]?

L: Did you people consider...or did anyone ever sit down and say, Gee, we ought to develop a little Cadillac?

B: No, I don't think so. But after it came out, why, many people did call it a small Cadillac. And, of course, the same basic principles had been used in the Cadillac engine: the striving for lighter weight and more efficiency. So it was a whole program.

We found a very capable engine designer in Chevrolet named Al Kolbe. Al had been supervisor of the drafting room, and he liked management. He was delighted to become the chief designer on this engine for me. He had worked on the Packard Twin Six [V12] way back when. He was an older man, now deceased.

And then his immediate assistant was Don McPherson, who's now [in 1974] GM's director of engineering at Cadillac. Those two really worked with the design crew and came up with, I think, one of the lightest-weight-per-output engines that's ever been developed.

L: Did the horsepower race raging at that time influence you greatly?

B: No. I don't think the horsepower race had hit Chevrolet at that time. The power race had already hit Cadillac and Chrysler and Lincoln, but...

L: ...it hadn't filtered down to the lower price ranges.

Also by Feb. 1954, a running engine very much like production could be fitted into front frame rails. Dual-action fuel pump with vacuum booster, scriptless rocker covers, and odd upper A-arm stamping are among this photo's anomalies.

Although shift linkage needed to be hooked up, and power steering pump was still wood, this engine compartment was starting to come together. Inner fender well was apparently hand welded. And for some reason, the Bendix power-brake booster was held in place by wires.

B: Actually, well, Pontiac, Buick, and Chevrolet were a little late with their V8s.

L: Yet when Chevrolet dived into stock-car racing in 1955, they did well right away.

B: Right, but that was after the V8 was introduced. I think [racing] was on our minds. [We had developed] a Cadillac that could move, and we ought to have a Chevrolet that could move, too; that would be agile and would give good fuel economy. Our little car succeeded in that.

But, of course, the big thing on this [V8] was the parallel development of a new engine plant in Flint: a tremendous undertaking. Right from scratch, and several times during the program, there were more lines added. It was always tough to keep up with the demand for that V8 engine after it was out in production. We finally had to open up and throw some of them at Tonawanda...that plant up there. At that time we were so high on V8s that we figured in 1956 that, boy, I tell you, we'll close down the last Six engine plant. The V8 proportion did get quite high.

L: Was there much overlap between the development of the 1955 Chevrolet V8 and the 1955 Pontiac V8? Or did those programs follow independent paths?

B: Well, there was only one overlap at that time. We knew they were going on a V8, and their concept was a heavier design than ours. But they had developed a sheetmetal rocker arm that we thought had possibilities. It hadn't been decided yet, but we jumped into that one and gave it to our manufacturing group, and they determined that they could make this rocker arm with no machining whatever. Just a metal stamping.

We also lubricated ours differently from Pontiac. We lubricated it with oil coming up through a metered tappet [hydraulic valve lifter], through the [hollow] pushrods, into the rocker arms, then over to lubricate the ball and the pallet of the rocker arm. The ball was mounted onto a stud that was pressed into the cylinder head. These were all new ideas and very good ones so far as automation was concerned. You never screw anything; just press the studs in.

L: When two GM divisions develop something like different engines concurrently, are they allowed to borrow ideas from each other? I don't understand how this works. In other words, why didn't Pontiac's 1955 V8 come out more like Chevrolet's?

B: In a way we were competitive, and in another sense we cooperated.

L: But you did know what was going on among your competitors?

Technicians seem to be making sure V8 can be lifted out of engine compartment for repairs. Vacuum-booster fuel pump was still in place.

In another fiberglass buck, engine was still mostly wood, including battery and exhaust manifold. Accessory oil filter had been fitted.

B: You've [always] got common knowledge, yes. Sometimes that [information] trickles down from the group vice president, who's pretty well part of what's going on in all the automobile divisions. He sees something he wants, he thinks it'll benefit the operation as a whole.... I don't know how they think, but I do remember visiting Pontiac in 1952 and talking with the engineers.

That stamped rocker arm...that single thing was one of the distinctive features of the Chevrolet V8 [that advanced] it over the Cadillac V8, because it enabled us to build the valve gear quite light. So far as inertia was concerned, this light valvetrain accounted for the higher speeds we were able to get when they found out this engine had racing potential. The engine would easily turn 5500 to 6000 rpm.

L: What about the development of the rest of the car? Was there anything radical or unusual about the program?

B: Well, the frame was entirely different. Chevrolet had been using a K-type frame—two channel irons and K sections. We envisioned a much lower rocker sill...trying to have the frame inside the rocker sills entirely. In fact, that was the first move that way. Almost all frames are that way now, you know.

We had quite a capable heavy sheetmetal and stamping group up in Flint. They came up with the idea of making the frame out of tubes. There's an SAE [Society of Automotive Engineers] paper on this frame, but they just obtained tubing...I don't know whether they rolled the tubing there or whether they purchased it from the steel mill [rolled], but the frame would start with about a 3- or 4-inch-diameter tube, and [they'd] bring it around in presses until they had a frame bent to the right siderail sections. They put in the right bends to miss the wheels, to attach the suspension parts to.... So this was quite an innovation and was, as far as I know, quite an economy feature in the manufacture of that frame.

And the front suspension was quite new [for Chevrolet]. It had an anti-dive feature....

L: The body became, as I understand it, an integral strengthening member of the frame. The frame itself was very light, wasn't it?

B: No, I wouldn't call it very light. It was not a light frame. I don't think it was lighter than any other, but it had much greater bending and twisting resistance. The body was what the corporation called the A-body. We shared it with Pontiac.

L: And development of the body...was that shared with Pontiac, too?

B: The development was mostly left to Fisher Body. The way the body's developed...in the planning concept [stage], Fisher Body Div. works with the customer divisions like Chevrolet and Pontiac. Fisher works with the divisional engineers and the general manager for the styles they want, and finally Fisher Body takes on the manufacturing.

L: Getting back to the engine for a moment, can you remember any particular problems in the development of that new V8?

B: Well, one had to do with getting proper lubrication to the rocker arms.

L: And was that the first time hollow pushrods had been used?

B: I can't swear to that. I think it was. Cadillac used solid ones. I forget whether Pontiac used solid ones or not. They might have used hollow pushrods just to save weight, but they took oil up through the studs.

Well, there were also problems with the hemispherical ball; the mating surface of the rocker arms. Also in the amount of oil that came up through the hydraulic lifters. One of our engineers, Papenguth was his name—he's over at Cadillac now—he came up with a little wafer, a method of metering that oil in the lifter. And that was one point that...a little logjam that was essentially overcome.

Then there was the mating surface between the rocker ball and the arm to be sure that contact was broken. That contact was heavier at the lower surface. In other words, [we needed] some mismatch of contact [in order for oil to get between the two surfaces]. And then [we had to have] the right amount of [oil] metering and distribution. [Too tight clearances caused some initial squeaks at the rocker balls in early production engines.]

L: How about the manufacturing end of the V8? Any problems there?

B: No, none that I can recall. It was an automated line and, as far as I know, it went without any great amount of difficulty.

L: Any other points?

B: I should mention that, even before the 1955 models were announced, we had a few people within Chevrolet who were proponents of more performance. The car came out with a 2-barrel carburetor and a normal compression ratio needing regular fuel. [Indianapolis driver and Chevrolet consulting engineer] Mauri Rose was one of those who first worked on the development of a Power Pack, which consisted of a 4-barrel carburetor, higher compression that used premium fuel, and a dual exhaust system. I forget whether that was a 1955 or '56 option, but the combination kind of set things on fire. [The Power Pack was used by] a fellow named Jim Reed, who raced a 1955 Chevrolet on short tracks and dumped everybody around the country. It was immediately picked up by the racing fraternity. Hot rodders worked on it with different cams....

L: You mentioned Mauri Rose. Was Zora Arkus-Duntov there too at that time?

B: Yes. He was. But I remember Mauri more right here because he was generally more of a tester type and Duntov was more of a designer type. When we came around to designing specific items, this first thing involved just the 4-barrel carburetor, as I recall. But later Duntov came up with different camshafts. This was for 1956 and '57.

L: When you originally laid out that 265-cid V-8 for 1955, did you have any idea it would go to 400 cubes and have such a long lifespan?

B: Oh, no. No, no. We knew we had room that was liberal. The cylinder walls were 7/32nds, as I recall, and plenty of water between bores, so as foundry practices became more refined, why the very next year we went to 283 cubic inches with just a change in the bore, which was minimal.

And then about that time, Duntov and the guys got to working on...Rosey [Maurice S.] Rosenberger got all enthused about fuel injection. The corporation was looking seriously into [fuel injection]. Rochester Div.

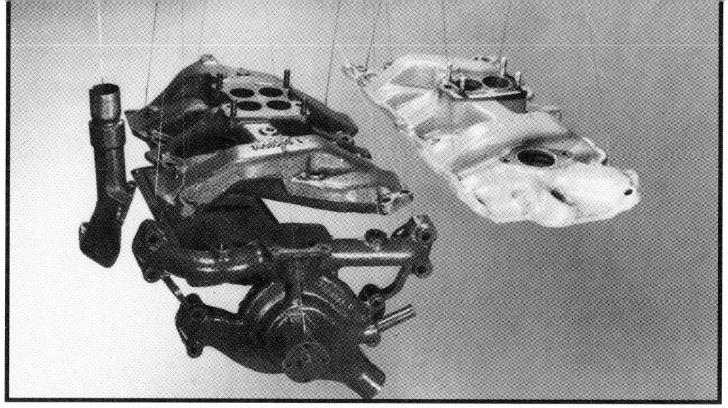

Cadillac intake manifold flanks simpler and lighter version from 1955 Chevrolet V8.

Forged small-block crankshaft combined rigidity with minimal overall length.

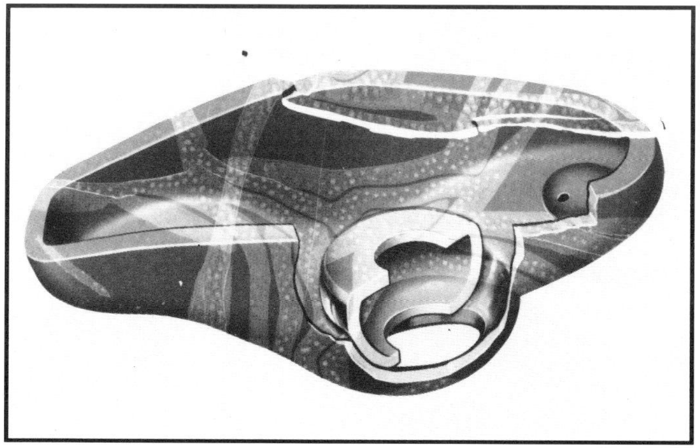

Clayton Leach conceived his masterpiece, the stamped rocker arm, in his kitchen.

Spoked flywheel and bellhousing adaptor were used with Powerglide transmission. The V8 engine itself was an option, and Warner Gear's overdrive also became available for first time in '55.

Finished chassis awaited a prototype body outside Chevrolet engineering center. Painted tailpipe indicates that engine hadn't yet run.

had developed one in competition with Bendix and others, and all [the suppliers] were submitting them to us. We finally selected the Rochester fuel injection as an option in the Corvette. In 1957, we made it available on the regular Chevrolet.

I remember Boss Kett [Charles F. Kettering] taking delivery of a brand-new black and silver 1957 Chevrolet Bel Air with fuel injection and then coming back later and reporting to us all the things he thought were wrong with it. He lived down in Florida in the wintertime. He came in and said, Ah, I have a list of things here on your fuel injection, and I want to talk to you. And he talked to the engineers for an hour and a half. He was enthusiastic about it, but that unit was before its time, and too expensive. It finally died out on the Corvette.

But the point I'm leading up to is that GM Engineering Staff would run tests on competitive and all GM engines, and the fuel-injected 283 was advertised as developing 283 horsepower—one per cubic inch. The guys at Engineering Staff said, Ah, no way! It'll never get that much horsepower! But they were amazed. They ran a test on that engine and showed up to 291 horsepower.

L: How big did you see the original 265 going in its lifetime?

B: We didn't actually see ahead of time where this engine was going. When you engineer a new engine, you allow some room when you start, but still you don't want to sacrifice the inherent weight advantage. We never conceived back then that we'd ever siamese the bores or go to 400 cubic inches.

L: And did you see the Chevy V8 staying in production as long as it has?

B: I don't believe we had any concept of that. A good engine ought to last 12 or 15 years. We redesigned the Cadillac V-8 in 1936 and again in 1949. That's 13 years, and it's considered about average.

Harry Barr

Interview date: 16 Jan. 1975, second conversation.

Lamm: In talking to Mr. Cole, I got the impression that he was originally asked to come to Chevrolet as manufacturing manager, not as chief engineer. But somehow he ended up as chief engineer. How did that come about?

Barr: Of course, Cole had been production manager at Cadillac [before he became Cadillac's chief engineer]. And then he went to Chevrolet in May of 1952. I got there two weeks later, in June, and he was already Chevrolet's chief engineer. So I don't know what went on between him and higher management, but I do know he felt more of a challenge in Chevrolet Engineering than he would have in manufacturing.

What happened: Ed Kelley had been Chevrolet chief engineer. He moved to manufacturing, so it might have been that they had an opening in manufacturing at that time.

L: But they didn't have an opening for a chief engineer, did they?

B: No, but when Ed Cole moved...see, Ed Kelley was chief engineer. Cole went in as chief engineer. They moved Kelley to manufacturing manager, see? So there might have been an opening there. I don't know. There might be something to your story.

L: Is the position of manufacturing manager considered higher or lower than chief engineer?

B: They're equal. And Ed Cole and Ed Kelley got along very well. Kelley, I remember, the first time I met him...had a big sign up on his wall. It said, "Eddie, anything you can design we can build!"

L: Now who came up with that green-sand casting method? I've heard that John Dolza at Pontiac had something to do with it.

B: John was always trying to sell something. I remember him coming down to the Cleveland tank plant one time with a car and a little V6 engine he'd been playing with. He let us ride in it. You know, he was in charge of engine development in GM Engineering Staff at that time. So he was always calling on Ed and me to promote ideas. You could be right [that he had something to do with the green-sand coring method]. But have you talked to Jim Premo yet?

L: No. I hope to soon.

B: Well, you see, I had responsibility for what was underneath the sheetmetal, and when you talk about the all-new 1955 Chevrolet, which was a remarkable little car, Jim Premo was my co-equal. He was the engineer who masterminded the body design. You really ought to talk to him about that. ❑

Ellis John (Jim) Premo

Assistant chief engineer, body, 1955 Chevrolet.

Jim Premo attended Cass Tech in Detroit and Chrysler Institute. He spent his entire career with GM, arriving at Chevrolet in 1935. His specialty was body engineering, and he acted for many years as liaison between GM Styling, Chevrolet Engineering, and Fisher Body Div. Premo oversaw the 1955 Chevrolet's body development in conjunction with Fisher Div.

When Ed Cole moved up to become Chevrolet's general manager, Harry Barr became chief engineer and Premo's title became executive engineer. After Barr was made GM's vice president of engineering in 1962, Premo took over as Chevrolet's chief engineer. He was later called upon to reorganize Opel's engineering department, after which he served as GM director of forward product planning. He retired in 1968.

Interview date: 28 Jan. 1975.

Lamm: As I understand it, you were an assistant chief engineer on the 1955 Chevrolet, on an equal basis with Harry Barr.

Premo: Yes. Harry handled the chassis. I handled all the body, electricals, and accessories.

L: Tell me about developing the body.

P: Well, the body for the 1955 model was engineered primarily by Fisher Body Div.

L: How did that work between the divisions and Fisher Body?

P: The division had responsibility for the underbody, the cowl, and the dash—things that have the closest relationship to the chassis parts. The instrument panel and things of that nature were up to the division. Then Fisher Body picked up the body upper structure in the interests of the corporation as a whole. That gave them a chance to do some integration between corporate divisions—to share bodies, like the A-body, which this was. That helped keep tooling costs down.

L: Chevrolet and Pontiac shared the 1955 A-body, but was it also shared with Oldsmobile the way it had been before?

P: No, Olds wasn't in on it that year.

L: How much did the body contribute to the strength of the overall structure? I hear the frame was relatively lightweight and lacked rigidity.

P: [Pause.] What's running through my mind is how these things you're saying will appear to people who read this. Actually I think the proper way to put it is that there was a very decided effort to join the capability of the body to the frame; to put them in relation to one another, so that each

Early sample bodies left Fisher plant in Feb. 1954 to be mounted on prototype chassis for road testing. One body still had split windshield (center). Doors were drilled above and below windows for trim but not along sides or rocker moldings.

With sample engines and bodies complete, track testing began in Apr. 1954 at the GM Proving Grounds near Milford, Mich.

added to the other in making up the whole. Together they provided a good, rigid structure. This was really how it was.

I don't even remember the exact dimensional relationships, but I do remember very clearly what our objectives were in those days. We were very aware of the need for a more rigid structure than we'd had in previous models. I'm sure there's an interesting story on that, and it might come out someday.

You know, there are two ways that a body engineer can go. You can go more rigid, which means that any vibrations you have are in the higher frequencies. These show up as harshness. However, if the structure is very flexible, it shows up in the lower frequencies, which is called shake. So rigidity versus shake is always the bugaboo. How stiff can you make it and how soft can you make it? At one time, there was an effort by a fellow in Fisher Body to make structures so soft that they simply wouldn't respond [this was in the interest of ride smoothness].

L: So any modern car body has to be a compromise between those two extremes?

P: I can't speak for the people who design cars now.... Personally I believe the ideal is to make the body infinitely rigid and then use isolation [to mitigate the harshness]. That's the argument for a separate frame, because it provides for isolation between the body and frame.

The thing that was so important in the 1955 model was that it was a whole new automobile. As Ed Cole mentioned, today it's almost impossible to design an entirely new car because of all the restrictions. The '55 Chevrolet was Ed Cole's big chance. He made the car and the car made him. He was a dynamo. He was very anxious that everything be good. Now Ed depended on me a great deal for the body. That was not his specialty.

L: But apparently Cole was very much interested in the design of the body.

P: Oh yes, always. But body structure at that time had not been his sport. The engine was his big dream, to get a V8 into Chevrolet, where we'd had only Sixes before. And that's what made that car. The big thing, I think, was its appeal to youth. Getting this car, a good high-performance automobile—this would take Chevrolet out of the doldrums, out of the old-man kind of business. You get the picture. ❑

Charlie Bidwell tanked up first running '55 prototype, #5001. Early mules used handformed '55 bodies with 1953 front clips and tail lamps.

Fifth wheel gave information on speed, acceleration, and distances. Old 5001 would eventually log 25,000 hard-driven miles.

First sample bodies didn't have the dipped beltline, nor did the 1953 Chevrolet front fenders match 1955 doors. Both bumpers were 1955.

John Dolza

Development engineer, 1955 Chevrolet.

John Dolza was born in Turin in 1902. He studied electrical engineering in Italy and came to this country in 1926 as a layout man with New York Edison. Dolza took a job with Buick the next year and worked himself up from draftsman to Buick's assistant chief engineer. He was put in charge of a GM engine research project in 1939 and, from there, went on special assignment to design aircraft engines for Allison Div. After the war he became involved in various aspects of engine research within GM central Engineering Staff, where he conceived, among other things, the green-sand casting method. He was also instrumental in developing the Rochester fuel injection system. Toward the end of his career, he held 120 patents. Mr. Dolza retired from GM in 1957 as head of the GM Technical Center in Warren, Mich.

Interview date: 24 Jan. 1975.

Lamm: Tell me about the green-sand casting technique.

Dolza: That was my patent. The development was done between myself and Pontiac Div. in connection with the development of a V6 engine. That was around 1947, the very first one, and it was done for Oldsmobile.

I was with GM Advanced Engineering, which at that time was called Product Development. I'd been transferred to Product Development in 1939 from my position as assistant chief engineer at Buick. During the war I was with Allison Div., until 1944, doing aircraft engines. In 1944 I resumed my work with Product Development, which later became GM Engineering Staff, under O.E. Hunt, and I was transferred to the GM Proving Grounds in Milford, Mich.

At that time Cadillac had developed a beautiful V8 engine. The author...although naturally there are always other people helping...but the person who deserves the most credit on the Cadillac V8, which became a turning point in automotive engine work, was Chris Bouvy. He should be credited with the original postwar Cadillac V8. It was a wonderful engine....

L: I had always heard that Ed Cole designed the 1949 Cadillac V8 himself, personally. That's the traditional view, I believe.

D: No, no, it was Bouvy. That's an unfortunate thing in big corporations. The top man gets the credit. Bouvy was the head of Cadillac's engine group at that time. Harry Barr and Ed Cole themselves attributed the 1949 Cadillac V8 to Bouvy. The boss gets the credit, though, unless he tells management about [the actual designer]. In fact, Chris Bouvy got disappointed with Cadillac and quit; went to another job in Chicago where he designed a group of V8 engines for industrial use.

Anyway, getting back to the 1955 Chevrolet V8, the GM central office

Flat-top hood accommodated new V8. Proving Grounds testing included several types of roads and artificial weather conditions.

Bidwell weighed prototype's front and rear to assure 1955 proportions. The reason for J.C.Whitney headlight bullets remains unknown.

To simulate different body styles and load conditions, sandbags were added to test car's trunk. Bumper cutout obviously wasn't final.

According to this Proving Grounds tracking chart, Chevrolet was running nine 1955 mules, five 1954 cars, and 13 1955 trucks.

realized that the V8 would be a success for the future and that it would replace the in-line Eights and Sixes of the various divisions. Since I was in advanced engineering, they asked me to develop a V6, because the V8s could just copy Cadillac. So I developed a 60° V6 for Oldsmobile. It was a wonderful engine for fuel economy and other things, but it was too expensive to produce because, just like Chris Bouvy's V8, the casting of the cylinder block was way too costly. The V8 required 24 baked sand cores, which in turn took a lot of steel rod reinforcement to lash all these cores together; a most expensive item.

Unless this problem could be solved, there was the possibility that in-line Sixes would never be replaced by V6s nor that the V8 would get into extensive production because of cost. Well, in the meantime, Pontiac got interested in a V6, which I designed and developed for them. Pontiac's general manager was Arnold Lenz, who at one time had been manager of foundries for Chevrolet. Lenz later got promoted to run Pontiac as general manager.

So I designed a V6 for which the 24 cores were reduced to four and didn't require any steel rods. I did the casting by turning [the block casting] upside down. But when I submitted it to the Pontiac foundry, they resented that I, an engineer who'd never run a foundry, would tell them what to do. So they said, Nothing doing!

Well, I went to Arnold Lenz, who I knew from Chevrolet, and explained it to him. Arnold, being a foundryman himself, knew I was right, so he said

Without correct inner fenders, test-car engine compartments ended up caked with dirt after running through water hazards and dust.

to me, Johnny, I'm going to take my coat off and roll up my sleeves, and we're going to go together to the foundry. Which we did.

Well, when a division general manager goes in and tells the foundry that this is right...that by refusing to cooperate they might make it impossible for Pontiac to use vee engines in the future...that they were interfering with production instead of helping...anyway, he gave them a good sermon. And so those fellows went to work and developed these [green-sand] core boxes.

Then later on, Pontiac asked me to design a V8 for them. On that one I was just doing consulting work because I had instructed their personnel in the basic design of vee engines. I did not design it in my own department. And while this was going on and the coring technique had been proven by Arnold Lenz and his foundry, Eddie Cole was made chief engineer of Chevrolet.

Now Chevrolet had designed a V8 under Ed Kelley, but what they'd done was simply to scale down the Cadillac V8. So the cost was almost the same. The only difference was that it had fewer pounds of cast iron. It was ridiculous and would have been nearly double the cost of their in-line Six. Which was unacceptable. It would have increased the cost of the car $100 or so.

Eddie Cole wanted a *new* engine, a V8, but he didn't have anybody on his staff capable of creating that engine so it would be economical. But he knew me from the V6 engine development, which I'd showed him while he

Block surface tested new balljoint front suspension and steering components. Brakes, rear axles, and driveshafts were also heavily stressed on this course.

High-speed cornering showed good weight balance and stability, and testers realized early that the V8 gave tremendous performance.

Another mule had 1955 front grafted onto 1953 body. These scenes by engineer Paul King show on-the-road V8 testing in summer 1954.

After putting on miles, engines were torn down and inspected in the motor room of Chevrolet's experimental shop building. Each part was meticulously miked and scrutinized for wear. A V8 is undergoing teardown in the foreground, with a Six on the stand behind it.

was still manager of the tank plant in Cleveland. I brought a car there, and we drove it together, and I proved to him that the V6 could be made.

So he asked me, Could you lend me two or three men who know how to design engines? Well, that was mostly my work. I had specialists in various parts: crankshafts, cylinder heads, things like that. So I told Eddie, You don't need two or three men; you need my entire department. I'll clean my slate of projects that can wait, and I'll get you an engine all designed in a few weeks.

L: Let me interrupt a moment. What was your title when you went over to work for Chevrolet?

D: No, no. I didn't go over. I kept the work in my own department. My title there was engineer in charge, which is equivalent to chief engineer. They called us engineers in charge to keep us separated from production engineering.

My projects were meant to go into production within 3-10 years. The divisions wanted to go into production in a couple of years—up to three years ahead—and GM Research was doing development up to 10 years ahead.

So anyway, that's the proposal I made to Ed Cole and to Harry Barr, his assistant chief engineer. And Cole said, Okay, John, you do it. And I told

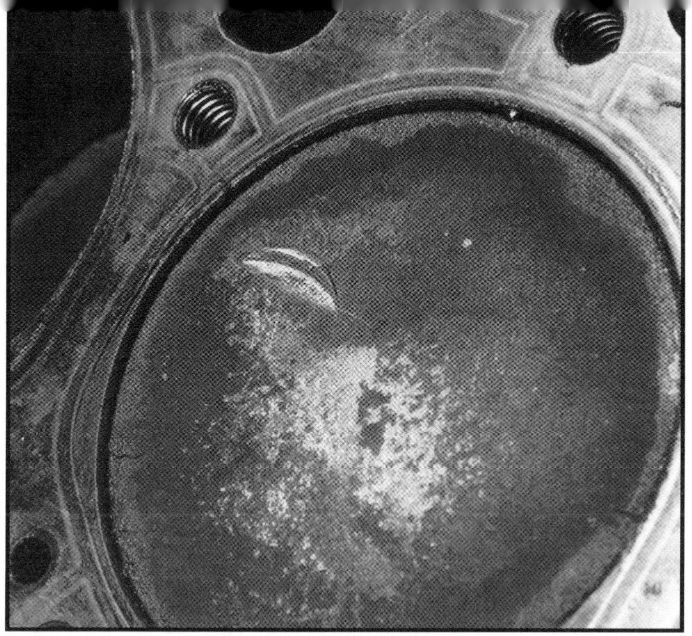

A few glitches did show up early in the engine program. Some rocker studs were set at the wrong height, making valves hit pistons at high rpm (top two photos). Rocker balls also squeaked until given proper clearance. Early V8s burned oil due to too-fine honing of cylinder walls. Some dealers poured Bon Ami down carb to cure this.

him, But you'll have to pay for this development. See, I never did anything for nothing with the divisions, otherwise they would dump on my department the projects that they didn't have enough faith in themselves, because then the GM Central Office would pay for it. So I said, Well, I'll charge you, and he said, That's all right.

Well, we worked day and night and weekends for about three months. We decided on the final configuration on a Friday. By Monday I had the bore-center distance—spacing of the cylinders. Also by Monday we had the people who produced the machinery to bore the cylinders to my place, and we ordered the base for those machines from just central line distances developed by computer over the weekend. Cole had that much faith in me.

We finished the project and introduced a V8 that takes five baked cores that make the crankcase, and I think Chevrolet added one more at the end. Altogether they put in six cores.

We also produced the stamped rocker arms. Those were invented by an engineer at Pontiac, Clayton Leach. This fellow had a very interesting background. He studied to be an astronomer, but he proved to be a very, very clever engineer. He, too, had the same problem that I had with the produc-

Chevrolet used nine major and three minor cores to cast 1955 V8, as illustrated on right (four cores not shown). This contrasted with 22 cores for Cadillac V8 (left). Thinwall castings meant less weight and metal but didn't allow much luxury in machining.

Foundrymen paste down pairs of cylinder-barrel cores as a first step in casting a V8. Photo was made in Dec. 1954 in Saginaw foundry.

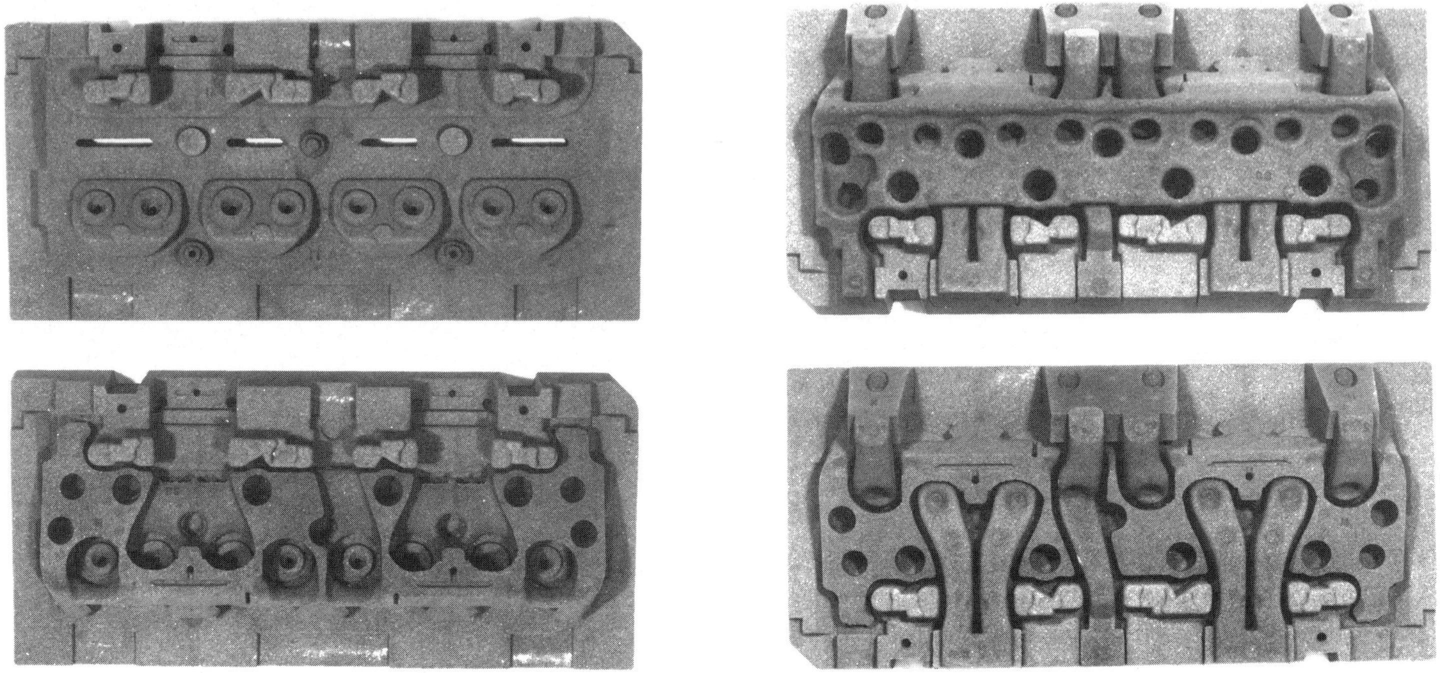

Four cylinder-head cores were used. The slab core included the four combustion chambers. Then the lower water-jacket core was set . . .

. . . on the slab. Next the gas-passage and upper water-jacket . . .

. . . cores were added, pasted together, and placed in the drag mold.

Two l-piece block jacket cores were slipped over cylinder-barrel cores and end cores were added. Cores were then dried in the bake oven.

tion department at Pontiac, and he made the first model of the stamped rocker arms in his basement at home. Because I encouraged him—he was my connecting link between Pontiac Engineering and my own department—I encouraged him because he could not get money even from his own chief engineer or production people to make a sample, so he did it in his basement, at home. I told him, That's right; that's good! Then, when he had a sample, we put it on the engine, period.

L: Was Clayton Leach the same person who came up with the hollow pushrod to lubricate the stamped rocker arm?

D: Hollow pushrods weren't new. The idea was very old. The hollow pushrod had been in production for at least 50 years. It was done mostly for weight reduction. So we developed the engine with those two features: the green-sand casting and stamped rocker arms.

L: Did you see that much potential for expansion in this engine? Did you ever think it would grow to 400 cubic inches?

D: No. It was built as a 265 originally, but it was laid out for 302 cubic inches, as I recall. The center distance that we ordered the machinery for was for a 300-cid V8. They went way past that, but they went past it by increasing the stroke.

Eddie Cole had asked for an engine with a 10-year lifespan. We knew the in-line Six was obsolete, but it had lasted from 1929 to well past 1955, so it was at least a 25-year engine. I told Eddie Cole, If we put some room for expansion in this new V8, it might last as long as the in-line Six. In my mind, I wanted it to last as long as the Six. Right now it seems it's going to do that. And I'm very pleased, because there's no other engine, not even Cadillac, that's lasted that long. Naturally the Cadillac V8 had to be redesigned to introduce my coring technique and other savings.

L: Let's get back to this green-sand casting business. Explain to me how it works.

D: Green sand is when you put the pattern into a foundry flask and just blow upon the damp sand and tamp it down to get an impression of the shape of the pattern. A dry core is when you blow the sand, which has a different binder, into a box, then either cure it slowly on a core mold or make it hard like brick.... The core is baked. The green sand is not. It's just like a sandcastle on the beach.

L: Did Clayton Leach have anything to do with the green-sand process at Pontiac?

D: He applied it to the V8 engine under my direction, but the original development applied to the V6, and the original man who helped me along was Arnold Lenz. ❑

Dried cores were set into drag mold, which formed outer walls of block casting. Dowels visible beneath cylinder barrels will become freeze plug cores. Sand mold for block deck was then set on top. Everything was turned upside down before molten iron was poured in.

Clayton B. Leach

Engine development engineer, 1955 Pontiac.

Clayton Leach was born in 1912 and grew up near Pontiac, Mich. At age 16 he scratchbuilt a running 4-cylinder aluminum engine. Leach received his B.A. in chemistry and math from Missouri's Park College and then went to GMI for three years. He began at GM as a draftsman for Pontiac, working his way up to senior project engineer in 1949, then assistant motor engineer in 1954, and Pontiac executive engineer in 1961.

Leach maintained a fully equipped machineshop in his home and developed many ideas there, including the Pontiac and Chevrolet V8s' stamped rocker arms. Leach involved his entire family in Pontiac test drives. According to his son, "Every trip began with a request for total silence to listen for deviant engine noises." In 1963, Leach became assistant chief engineer for Adam Opel AG in Germany, then was transferred back to Pontiac in 1966, and became Pontiac's assistant chief engineer for chassis and powertrain in 1970. He retired in Jan. 1973.

Interview date: 22 Jan. 1975.

Lamm: Apparently the 1955 Chevrolet V8 had some of its roots in the Pontiac V8. People I've talked to suggested I get in touch with you. They mentioned that you held patents or were responsible for the oiling system,

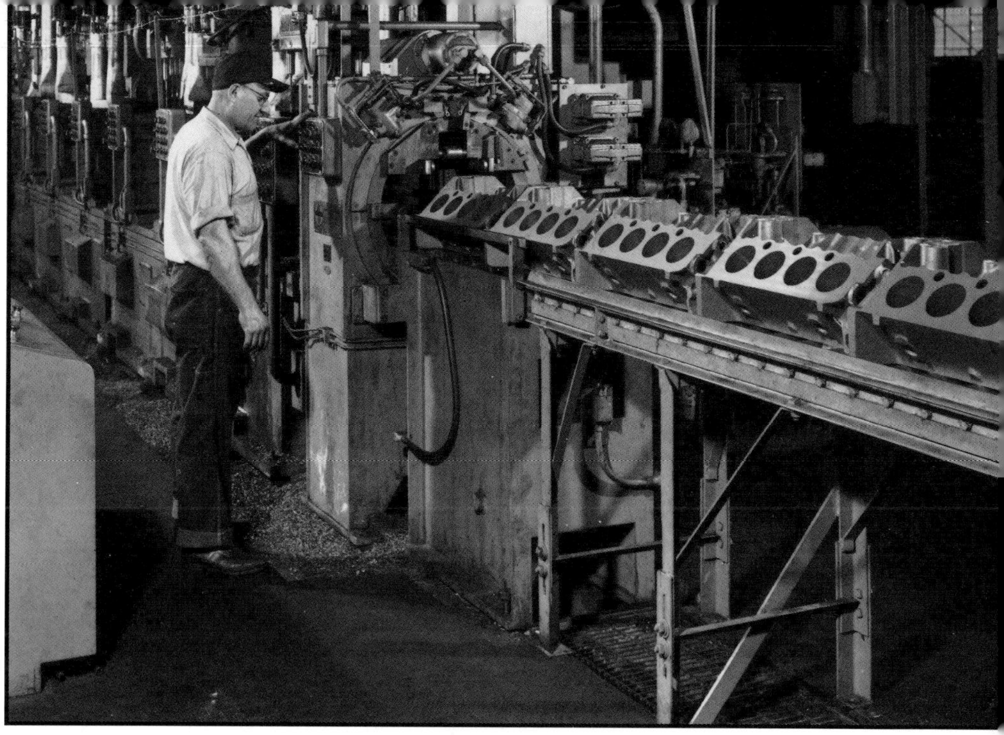

Preliminary machining of block and cylinder bores took place here in the Flint engine plant as soon as castings had time to cool.

Just-cast engine blocks rolled through rough broach at far right. This machine planed deck surfaces before shipment to machineshop.

Technician sized cylinder bores on both banks and marked each one with a number so piston installers could match pistons to bores.

the hollow pushrods, stamped rocker arms, or at least that you were in on the development of those things.

Leach: That's right.

Lamm: How far back did that go?

Leach: You mean the development of those items? I conceived the idea and made the initial parts in my shop at home in the basement.

Lamm: When was that?

Leach: That was just after World War II, about 1947. We started the development on that V8 engine at Pontiac just before World War II and then, of course, we had to stop [for the war]. We started in 1940 and had to stop, and we picked it up again in 1945 and brought it up to initial production [?] in 1951-52. Went into auto production in the summer of 1954.

Lamm: Ed Cole mentioned that Mr. Papenguth developed a little wafer that went into the oiling system for metering oil up into the pushrods. What about Papenguth's wafer?

Leach: Yeah, Chevrolet had a different system. The wafer was in the hydraulic lifter. I know Bob Papenguth. He's a good friend. In Pontiac's system, the oil went up to lubricate the pivots between the pushrods and the rocker arms. Chevrolet didn't do that. Yes, they did do that, but then they squirted oil to lubricate the main ball pivot. Pontiac had pressure to the main ball pivot, so the oil that went up the hollow pushrod was trapped in there. It didn't escape. It didn't squirt all over.

So Chevy, in order to prevent too much oil from flying around, going down the exhaust valves into the combustion chamber...Chevrolet decided to save some money by not drilling the passages to the main ball pivots, which carry, of course, the main load. And when they did that, they had to get some lubrication on that, so they had to drill a small hole in the pushrod socket in the rocker arm, if you're following me.

Now later, Pontiac did the same thing, sometime in 1965. They dropped that pressure lubrication to that main ball, which, in my opinion, was a big mistake. It caused a lot of problems. They went back and did what Chevrolet did and lubricated everything up through the pushrod, and then you have to have that little disc in there to keep too much oil pressure from coming up. That's the control thing. If that's the one we're talking about.

Lamm: I ought to talk to Mr. Papenguth about that.

Leach: I say he's a good friend of mine, but I can't be sure what he contributed specifically. I don't recall that that clearly. I know he contributed to engines; he was a Chevrolet engine type.

Lamm: What was your title at Pontiac at that time? And how come you were working on it in your basement and not at Pontiac?

Leach: Well, I worked at it both places. I was working about 24 hours a day. That's the way you get ahead, isn't it? I had a real keen interest in...well, it was a matter of pride. You get an idea that's radical, and it stays with you. It looked like a piece of 10-cent-store tin, you know, but....

So to work out the bugs and make sure something was feasible, I'd often do this [work at home]. I liked to do it. And I didn't have time at work really to think about these kinds of things because I was so busy.

To answer your question, I was motor development engineer at the time. Well, I was a designer and then a project engineer assigned to the development of the new engine. In fact, that's about all I worked on for about 10 years. Then I became assistant motor engineer, then chassis engineer, then executive engineer, then assistant chief engineer at Pontiac.

Lamm: The other thing that made the Chevy V8 different was the green-sand core.

Leach: That was developed by Pontiac. Pontiac developed all the casting techniques, especially for saving core sand. I think we saved...the conventional way to cast a V8 block, I think, had about 350 pounds of dry sand. We had, I think it was, about 158 pounds. I can't remember the exact figure, but it was a very appreciable reduction.

Lamm: Who was the man who developed that?

Leach: John Dolza got a patent on a cylinder block for the coring. And actually this was a development that Pontiac was working on at that time for a V6 engine.

Lamm: Dolza got that patent for the V6?

Leach: I think we were working on that, and I think he got the patent. I think it's pertaining to V8s. It was kind of a mixed-up deal. At that time the patent section wanted me to be co-inventor with Dolza, but I said, Oh, I'm not interested in holding patents.

As far as the two engines were concerned, the Chevy V8 and the Pontiac V8, the best records would be the two SAE papers that were given in January of 1955. They were given together, and they really spell out the differences between those two engines. ❑

Bores received pistons here at this station, with pistons numbered and arranged by size on table in foreground. Each piston loader matched sizes and installed pistons along with connecting-rod assemblies. Overhead conveyor continuously replenished piston sets.

The inspector on the right used a dial indicator to check bellhousing concentricity. Main bearing caps can be seen already in place.

Malcolm R. McKellar

Motor development engineer, 1955 Pontiac.

Born and reared in Pontiac, Mich., Malcolm McKellar became a Pontiac man through and through. Developer of the legendary McKellar camshaft and several experimental overhead-valve and overhead-cam engines, including V8s, he initially joined Pontiac Motor Div. as a GMI student, graduating in 1941. After overseas combat service during WW-II, McKellar returned to Pontiac in 1946 as a senior detailer. He was promoted to senior design engineer in 1949, became an assistant motor engineer in charge of design in 1955, and motor design and development engineer in 1961. McKellar was promoted to staff motor engineer in 1965 and retired in 1982, after 45 years of service to General Motors.

Interview date: 22 Jan. 1975.

Lamm: I've talked to Mr. Cole, Mr. Barr, and a lot of other people, and the consensus seems to be that a lot of the innovations for the 1955 Chevy V8 came from Pontiac: the valvetrain, the hollow pushrods, stamped rockers. Were you in on that?

McKellar: Yes. Actually Clayton Leach was the originator of that system as it was developed.

L: And when would that have been?

McK: Well, I was....

L: I believe work started on the 1955 Chevrolet around May of 1952.

McK: I don't recall the exact dates, but I believe it was in the late 1940s that the first stamped rocker arm was made up by Leach.

L: And was that a GM Research project, or was it in Pontiac Div.?

McK: No, that was strictly Pontiac.

L: The other thing about that engine was the green-sand-core in the valley, the upside-down casting technique. Was that a Pontiac development, too, or Chevrolet, or where did that come from?

McK: I don't really know who holds the patent on that. I think the concept was first applied to a V6 engine with a 123° bank angle. I believe John Burnell probably got a patent on that. He was with GM Engineering Staff at the time. That goes back to the late 1940s, too. But then the application of that to a V8 was, I guess...I'd have to say pioneered again by Clayton Leach.

L: Apparently one of the problems was that the oiling system put too much pressure to the rocker pivots. Mr. Cole told me that Mr. Papenguth, whom I haven't talked to yet, developed some little wafer that made the oiling system practical.

McK: I don't know who developed that, but yes, there was a little flow

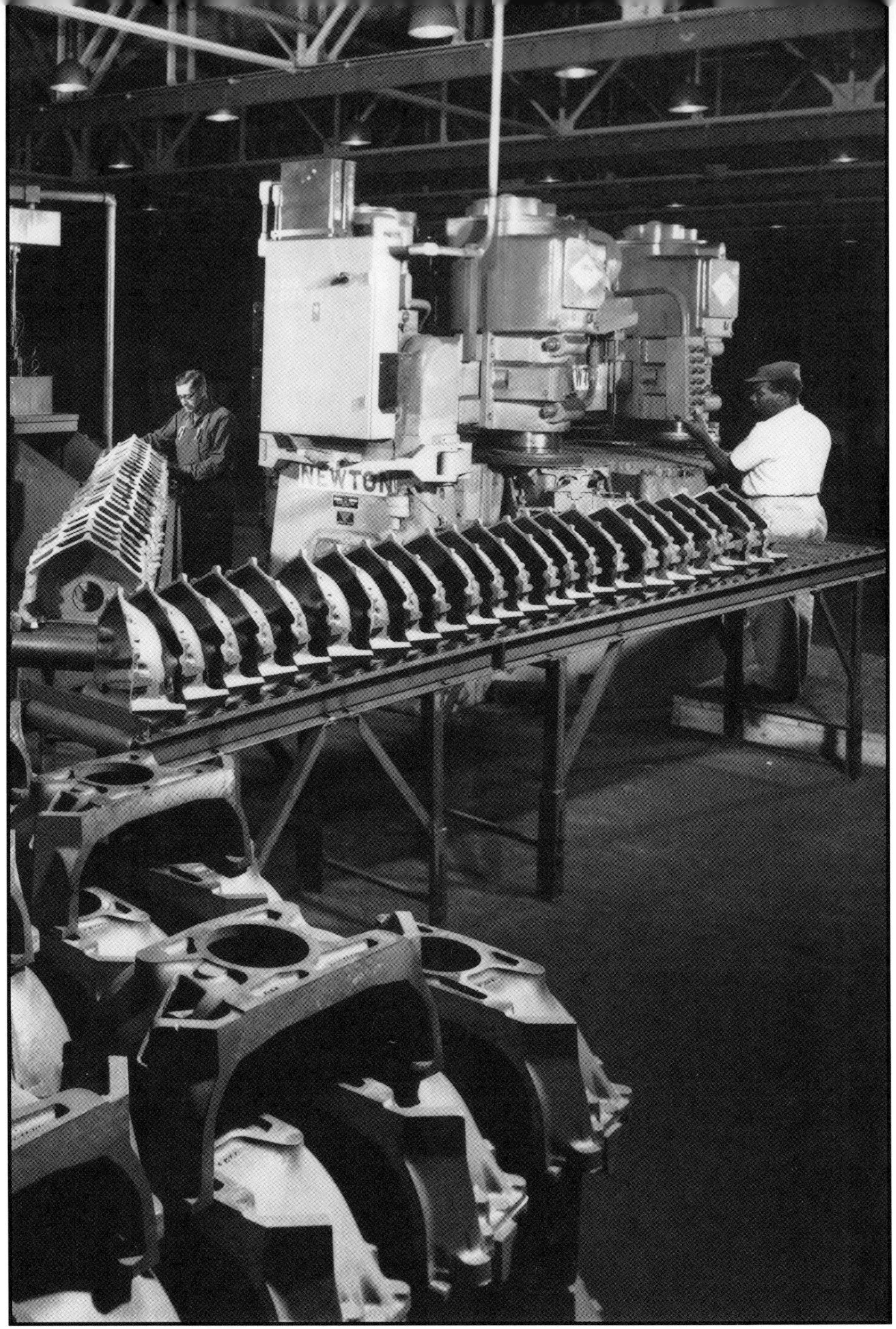

Before assembly, bellhousings passed through a 2-stage milling area in Flint engine plant where mating surfaces were trued.

Micrometers, gauges, and checking fixtures arrived here and were logged in before being sent to various areas of assembly plants.

Each freshly assembled engine was put on a dynamometer and run for several minutes on natural gas. Instruments checked for vibrations, tested oil pressure, and suggested carb adjustments.

Early test engines were periodically disassembled and inspected for wear, then reassembled and run more, either in a car or on the dyno.

restriction in there that, in effect, meters the oil. As you say, it's a little curved...not a disc...maybe a disc....

L: Mr. Cole called it a wafer.

McK: That's a good word for it. When the lifter moves up and down, that more or less opens up the passage for the oil to flow. We didn't use that for quite a while here at Pontiac. We had another way of doing it, and we finally ended up....

L: Were these things developed cooperatively, hand in hand, or competitively? Or how does that work? Do these ideas cross divisional lines?

McK: Well, I think what happened was that...of course, our engine was presented to corporate management, and they were shown the features and, because of the uniqueness of the valvetrain, Pontiac was told to give this information to Chevrolet. And Mr. Cole, who was then in charge of that project, had it incorporated into that Chevrolet design. This, of course, was in the early 1950s.

Of course, we had our own version of this valvetrain, and after our

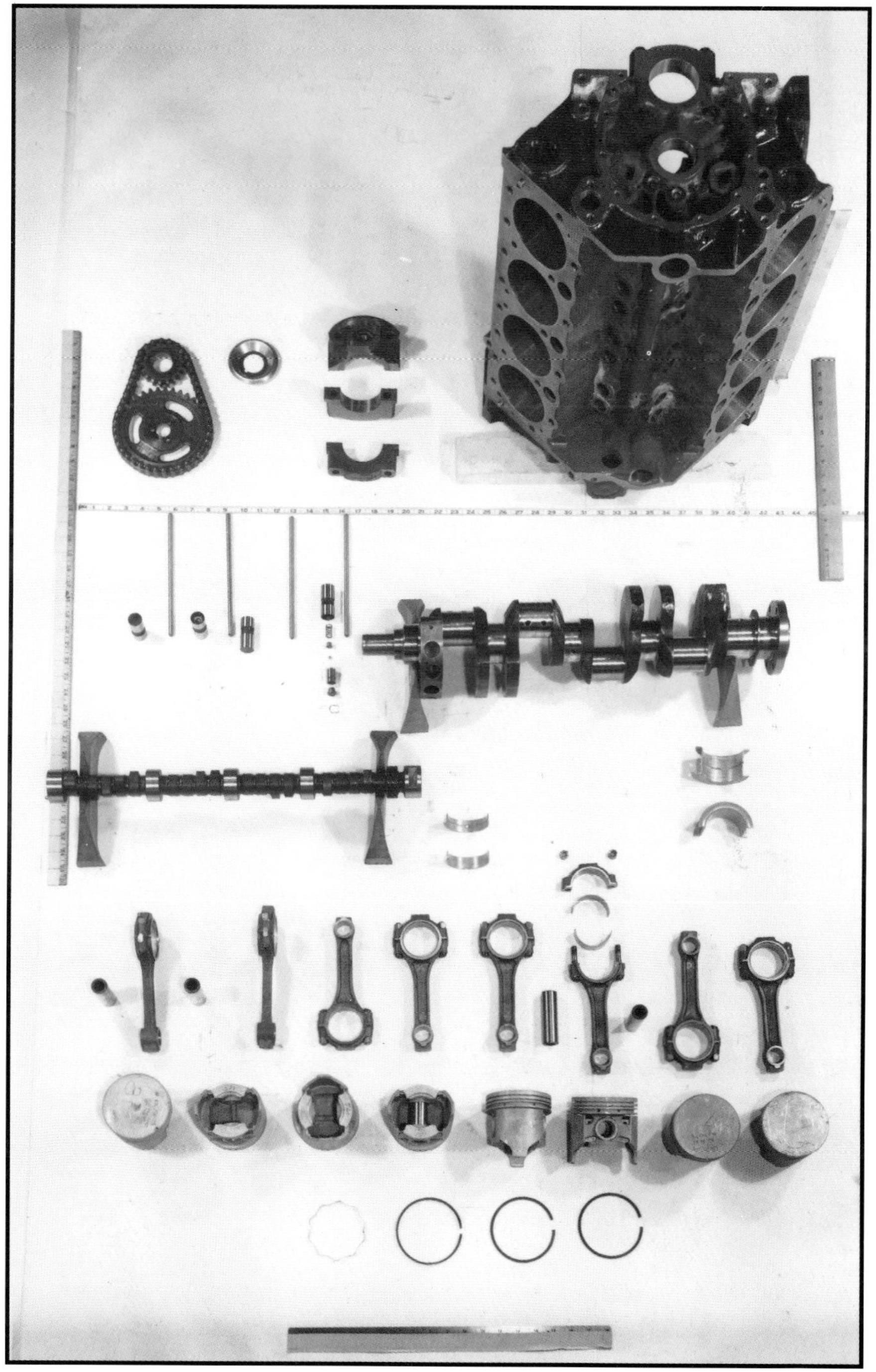

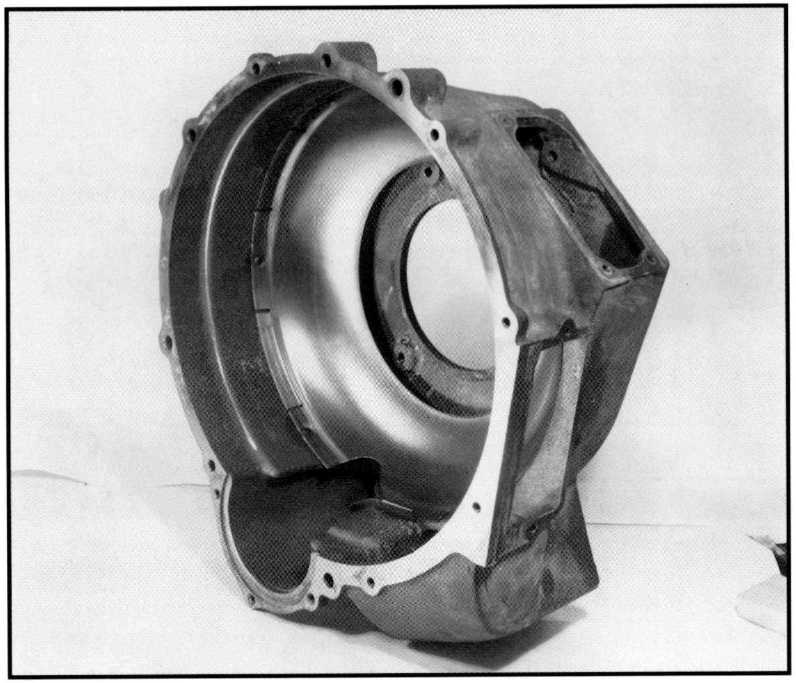

Bellhousing for adapting V8 to Powerglide transmission differed from that used for the Six by eliminating ear-like projection at starter.

Tests of optional Delco electric wiper motor showed failure after 137 hours. Internal switching mechanism was changed as a result.

design was turned over to them, then they chose to go their own way, and eventually we went our way.

For instance, right now we're different in that we [Pontiac] have a fixed pivot. In other words, the rocker-arm ball can't be adjusted up and down. We come down to a fixed position with a nut on a taper on the stud, whereas they [Chevrolet] have to adjust theirs manually. That's one of the differences.

L: And those were things the divisions worked out for themselves.

McK: Right.

L: What I'm not clear on is how much of the development work was done by GM Research, how much by GM Engineering Staff, and how much by the divisions. Specifically, where did that 1955 V8 engine come from initially? I know that Mr. Cole worked on the Cadillac V8. I don't know how much of that engine was all Cadillac, how much was Engineering or Research or Oldsmobile, but I was wondering.

McK: All I can say is that in the case of the valvetrain, neither GM Research nor GM Engineering Staff had anything to do with it. What I really think happened...I think Mr. Cole had all these ideas pulled together into what turned out to be the 1955 Chevrolet V8 engine, using the best concepts available at the time.

L: Did Pontiac also use that green-sand casting technique for its V8?

McK: Oh, yes. In fact, we had fewer cores to manufacture our block and head than anyone in the industry, and I think that's still the case. I wrote a little paper on that at one time that got published in the General Motors Engineering Journal.

L: How many cores was that? I know Chevrolet was very proud of the fact that they used only 21 cores for the block and both heads.

McK: I can get out my SAE paper. Hang on a moment. Well, I can't find it, but I can count them up. We had six cores for the block and four for each head. That's a total of 14, so that's seven fewer. Chevrolet didn't go quite as far as we did in simplifying the castings, but they did do some of the things that we used.

L: Another thing I remember Mr. Cole saying is that they really put their production people on their ears, because they cast down to about 7/32nd inch, whereas before 7/16ths was about the limit. So this was new to the foundry people. Was that the same at Pontiac, those sorts of tolerances?

McK: I'm not sure what he.... Those numbers don't sound right.

L: Let me get out my transcripts. It was 5/32nd that he said.

McK: Yeah, that would be 0.156 inch. Well, we didn't go quite as thin as that. Our spec was 0.180 inch. And, of course, whether they actually came out that thin for Chevrolet I don't know. That sounds a little bit on the thin side on the average [chuckles]. But anyway, I'm sure that this technique did allow more accurate setting of the cores, fewer cores, and thinner walls, so I'm not going to argue the point.

L: Well, thank you, Mr. McKellar. I appreciate the information. ❑

Handbuilt 1955 prototype, with numbers visible on floor panels, was upended to check weld sites, body and suspension attachments.

Test car #5203, a sport coupe, was photographed top to bottom in the Proving Grounds garage before embarking on its 25,000-mile run.

Same car, different view. No 1955 wheelcovers were available in Feb. 1954. Exhaust system showed considerable evidence of hand welding.

Suspension engineers cleaned up finer points of front-end geometry (top) while Fisher people figured ways to make body more efficiently.

Alex C. Mair

Executive engineer in charge of 1955 Chevrolet truck bodies.

Alex Mair grew up in Flint, Mich., where his father taught tool and diemaking for GM. As a student at General Motors Institute, Chevrolet was Mair's sponsoring division. He served in the Navy during WW-II and, in 1946, rejoined Chevrolet as an engineering draftsman. He became involved in truck engineering, quickly worked his way through the ranks and, in 1965, was named chief engineer of passenger cars and Chevrolet director of engineering in 1966. Mair became general manager of GMC in 1972 and Pontiac's general manager in 1975. He retired as GM's vice president in charge of the technical staffs group in 1986.

Interview date: 5 Jan. 1991.

Lamm: Let's talk about the development of the V8. I understand you were heavily involved in the 1955 Chevrolet truck program.

Mair: Yes, I was. The development of that engine began in the middle of 1952, and the V8 went into production in July of 1954. So it was under development for just under two years.

L: Now Mr. Cole got there in May 1952, and Harry Barr followed in June. But I hear that Ed Kelley was working on a V8 even before Ed Cole got there.

M: That's right. But then you have to remember that Chevrolet always had [various] types of engines under development. Once in a while, occasionally, it was a V8, but the most significant of the V8s was, in fact, the one that you just mentioned, that Ed Kelley, who was chief engineer at the time, had under design and some development around 1951.

Now the 1949 Cadillac and the Oldsmobile Rocket engine had been introduced. Studebaker and Ford, of course, already had V8s. Then Chrysler introduced the hemi for 1951. So the V8 engine was coming about in this world.

I can't remember the displacement of Kelley's V8, but it was relatively small—in the 230-cubic-inch range. I remember little things about [that engine]. For example, there was nothing particularly unique about it. It was fairly much like the Cadillac or the Olds engine, which were unique for their 1949 introduction, but by now it's 1951.

L: Did it seem to pattern itself on the Cadillac or Oldsmobile V8?

M: Yes. Because those engines existed, obviously Kelley and his people were looking at them. There were characteristics of each of those two engines. The Cadillac, as you may remember, was 331 cubic inches and the Olds was 303.

L: Did Ed Kelley's Chevy V8 use the same type of valvetrain and so forth?

In Nov. 1954, one of the first factory-assembled 1955s arrived at service garage for a production check and then further testing.

M: It was the same: center camshaft, pushrods, rocker shafts like those two GM engines. And then, around May 1952, Ed Cole came to Chevrolet, and at that time, I was the production engineer of an aircraft engine in Tonawanda, N.Y. Chevrolet was building an engine for the Korean War, an 18-cylinder radical aircraft engine—two rows of nine. It was designed by Wright Aeronautical and called the R-3350. I was the production engineer, and I remember distinctly, because I took a telephone call from Ed Kelley, who was indirectly my boss at the time; I took this call, and immediately Ed Kelley told me he would be the general manufacturing manager of Chevrolet, and Ed Cole would be the chief engineer.

L: Did that news come as something of a shock?

M: Well, I didn't have any idea it was going to happen.

L: And it was Kelley himself who called you?

M: Yes. The reason I took the call was because Hal Flynn, the chief aviation engineer, was out that day, and Ed Kelley called us directly.

We were viewed at that time as sort of a crack team working on this airplane engine, but Kelley said, You're all through with that now. Come back to Detroit. So I had one day's notice that Chevrolet Engineering was going to be all reorganized, and I must tell you that in all my life's work—in all the time I worked at General Motors—it was never so obvious that a new person was in charge...almost, I would say, one day after Cole got there, Chevrolet Engineering had a whole new feel. It was a vibrant feel, exciting, and you knew Ed Cole was there. Boy, I'll tell you: What a change!

One of the things he did...apparently he took a 5-minute look at the programs that were going on, all of them, and just said, We're not going to do any of these. And that was true. He started with an absolute clean sheet of paper for a 1955 line of automobiles *and* trucks. And *trucks!* They were to be all-new.

His plans included a brand-new V8 engine, because he had some ideas, and I will tell you quickly what they were. They were quite outstanding for the time. One was thinwall iron castings. Nowadays they don't look very thin, but they were for those days. This engine marked the breakthrough to thinwall castings.

Another innovation was to not have a separate stamped valley cover. Cole used the inlet manifold for that. The inlet manifold bolted directly to the heads and also became the valley cover.

Another big breakthrough was to get rid of the old rocker shafts and use

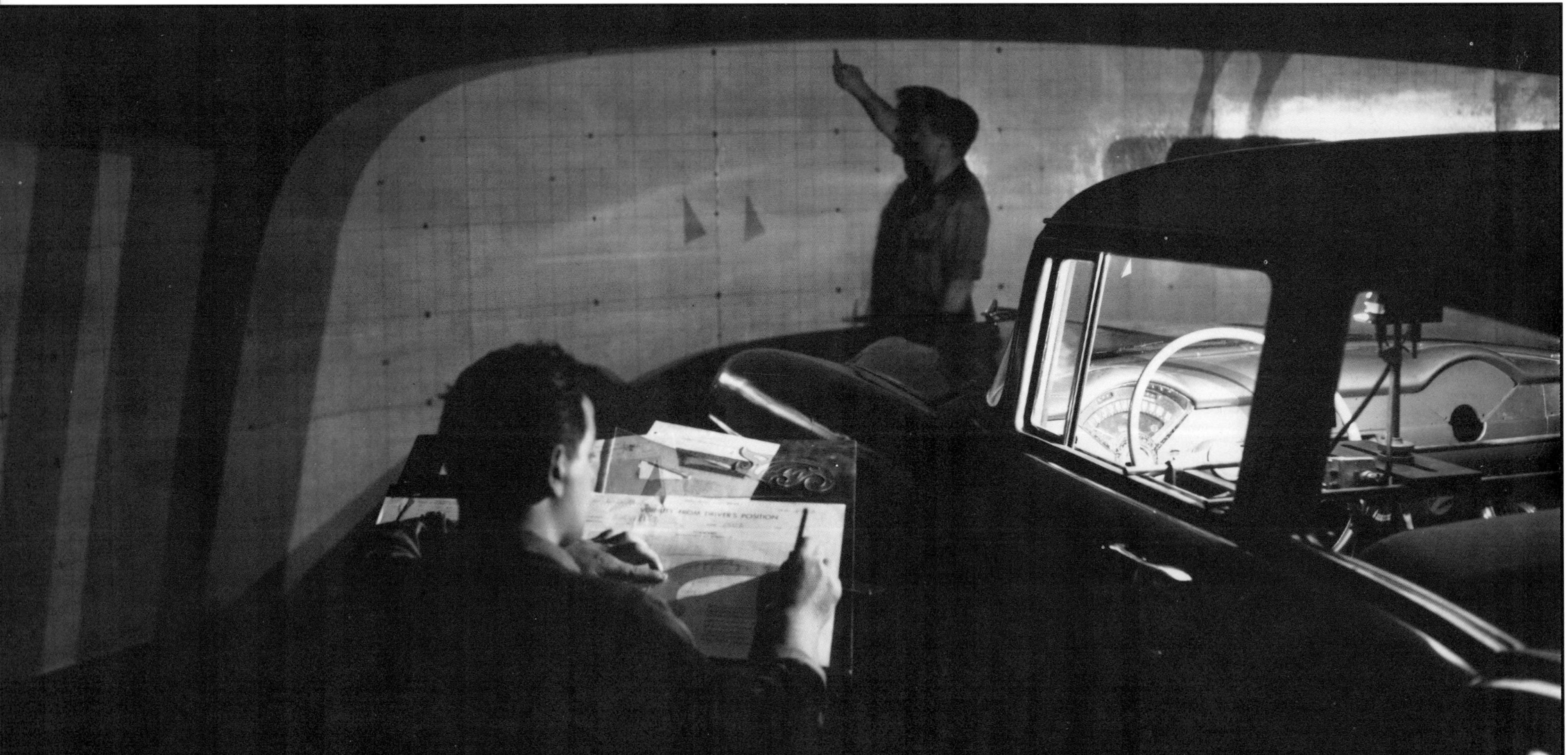

Wraparound windshield upped glass 19.4%. Check of blind spots mounted two lamps at driver's eye level, plotted shadows on wall grids.

Convertible chassis required a central X-member. This early prototype frame shows an extra brace near starter that ultimately wasn't used.

Fisher's sample convertible bodies, destined for Proving Grounds and road testing, weren't trimmed but had all major components.

a spherical ball at each rocker. That made all the valves independent, so wear or misalignment at one wouldn't affect the rest.

Then, too, we had an extremely large bore-to-stroke ratio. The 1955 engine had a 3-inch stroke and a 3 3/4-inch bore, which made it 265 cubic inches. So the engine had good high-speed characteristics for its day, and it was lightweight because of the thin castings and the short stroke.

L: How long was it after Cole got there that he had these ideas in metal?

M: I believe it was only about 15 weeks after Cole sketched up the ideas that the first prototype engine was built. Fifteen weeks! And as you know, it was barely 15 months later that this engine was in pilot production.

L: Was that one of the reasons they put Cole into Chevrolet, because he was so much more in favor of a V8 than Kelley?

M: There was more to it than that. Chevrolet had become so powerful and was so far ahead, outselling everybody...because almost everything Chevrolet did sold well. As a result, some of the top people at GM began to notice, and the sharper ones realized that Chevrolet was standing on its laurels. The division wasn't leading the way as it had in the past: the new 6-cylinder engine for 1929, all-steel roof for 1934, independent front suspension in the lowest-priced cars for 1935, automatic transmission and hardtop convertible for 1950; that sort of thing.

Anyway, they figured they needed somebody who would come in there and just knock the place apart. And Ed Cole did that, because as you know, Chevrolet sold 1,830,000 cars that first year of the V8, 1955, plus 393,000 trucks.

L: I had also heard that when Cole was hired, he wasn't offered the position of chief engineer at first.

M: He was offered the manufacturing job, which Kelley got, and he [Cole] refused it. That's true. I think they asked him to come in and put some fire under Chevrolet, and he said the only place you can do that—and at that time it was true in General Motors—was in the engineering function, where you controlled the product. Now Manufacturing could do some nice things, but they could hardly say, Let's do a new V8 engine. Well, *what* new V8 engine?

So I think that Cole's point was that if GM really wanted him to have this effect, it had to begin by giving him charge of the engineering department. So they proposed that to Kelley, and I always heard by the grapevine that Kelley accepted it quite properly.

L: What influence did GM Research or central Engineering Staff have on the V8 that Kelley was working on?

M: You hear various things but, in truth, there were always ideas flowing from the [corporate] technical operations into the divisions. The divisions usually looked at these ideas with jaundiced eyes, because the originator didn't have to take the risk. The final person who puts it into production takes the risk [responsibility for the idea's success or failure].

But there was always work going on [in GM Engineering Staff], combustion in particular. Also materials, like valve surfaces, things like that. But the actual design, I don't remember any real amount of [outside] work going on. Cole just said, Here's the way the engine's going to be.

L: But didn't he pick up these various bits and pieces from different sources?

M: You mean the parts of the engine? Oh, I think so.

L: Yes, like the stamped rockers and the hollow pushrods.

M: With pieces like that, there had been work done at the technical staffs

This was test mule #5251, equipped with V8 and Powerglide.

on how to make such a stamped rocker arm and how to sinter the iron to use for the ball. In fact, at one time there had even been a semi-cylindrical piece with a cylindrical thing stamped in, and I think some engines have that today, rather than a sphere. There had been work done on sintered balls and sintered bearings.

Then, on thinwall iron castings, there had been work done on flow rates into the sand cores, fewer pieces, etc. That's true. There was nothing inventive in that sense, but Cole brought it all together.

In those days at Chevrolet Engineering, though, it was a little different than now. You knew almost everything that was going on. There were hardly any meetings and almost no paperwork. Everything was [created] at the drawingboard, in the experimental shop, or at the factory. You were a part of all the things going on, particularly the engine, because it was so interesting.

Especially when they fired the first one up. Everybody was around looking things over. And then to have to use it in the pickup truck, as we did, and also in the medium-duty trucks the year after [1956]. We had to do the work, like put chrome piston rings in it and hard valve seats, like that, so it could go in the truck.

L: Tell me about when you first fired up the prototype V8. Who was there?

M: When I say they fired it up, I probably saw it the day after. I know that Ed Cole and Harry Barr and Rosey Rosenberger and Max Roench certainly were there when they fired it up. Anyway, I remember the second day, when I went to see the engine on the dynamometer, and for its day it had incredible pre-state acceleration because of the short stroke. If I remember right, it would accelerate at least 5000 rpm per second. It was the snappiest production non-race engine I'd ever seen. And it acted that way in the car.

L: I asked Mr. Cole, when he was alive, whether he had originally envisioned that engine going as far as it did in competition and displacement, and he said yes, that they had had that in mind.

M: It was true. Here's a little story I remember. It was 1969, and I was Chevrolet's director of engineering—the biggest engineering department in the world at that time. I came in to work one morning, and we were supposed be there at 8:30, and I usually came in at 8:15. When I came in that morning, there was a note on my desk to call Ed Cole.

Meanwhile, one of the draftsmen came running into my office and said, Gee, Ed Cole was here at 7:00, and he wants us to make a 400-cid engine out of the small-block. I said, A *what*? He said, A 400! He said he wants you to call him.

So Ed was a really nice guy, and I called him, and he said, Hey, where were you? I said, What do you mean? He says, I was out there and you weren't there. I said, Well, Ed, I hear you were here at 7:00, and we don't start until 8:30. If you want me to start at 7:00, I'll start at 7:00....

He said, I got you guys started. You ought to run up to the drafting room. We're going to make a 400 for *1970*. Nowadays, the modern world has slowed down. I don't understand it, but if they said that at General Motors now, they'd say, *When*?

But this was not even a year earlier, and Cole said, We've got to move on this. So I said, Ed, there isn't room for the bore in there. He said, I already did it. I sketched it up. We're going to siamese it.

And I laughed at that, and he says, Now Al, you already had personal experience in the truck business, and you siamesed the 235 engine, made a 261 out of it. And we had. He said, How was that engine? I said, One of the best we ever did. He said, Well, do that on the V8.

So I went up to the drafting room, and he'd already had the guy make a few hand-drawn sections, and we started right then. We called the foundry, as you could in that day, and they were down that afternoon from Saginaw, and we went into production in 1970 on a 400; came out 402, if I remember correctly. In fact, I owned a few of those engines in a marine business I was in for a while. They operate fine right now as marine engines.

L: Any other anecdotes like that that you remember?

M: Well, you might have heard that the initial reaction to the 1955 Chevrolet, strangely enough, was not fantastic.

L: Public reaction? Yes, I'd heard that.

M: The first two weeks after introduction were really slow. The car wasn't selling. So Red [Harlow] Curtice, who was then president of General Motors, got hold of Cole, and I was with the group in there, and Curtice said, The trouble is, we've got a grille that looks like a Ferrari, and Ferraris sell in the hundreds, not in the millions. So there was a crash program in

Some interior hardware, like horn button and clock, was still unavailable in early 1954. Hood-to-fender fits needed work, too.

What fun! Up-ended convertible eased shop-manual photography, but then one of the guys wanted his picture snapped sitting in the car.

Windshield-header hardware and top-fit adjustments for the 1955 convertible got a tryout on early Fisher sample body.

Styling to quickly change the grille. It never happened, because by that time the car began to sell....

L: And I'd heard that the early production V8s had problems burning oil.

M: That was just in the first few days. There were actually three items on the engine that they had trouble with. One had to do with the machine [on the engine assembly line] that pressed in the rocker studs. They had an automatic system in the plant which, if the studs were pressed in too high, the engine would be rejected. And if they were pressed in too low, the engine would also be rejected.

Anyway, somebody apparently loosened up that automatic system in the plant and didn't tell anyone. It was not rejecting the low end of the force being pushed in. Quite a few of those engines, when the driver would over-rev them—and the engine was so lively that people weren't used to that—they would run up to the valve-float speed, and the valves would touch the pistons when they stayed open too long. When they did that, the force would push the stud out of the head a little ways. And, of course, then when it came down, the valve wouldn't open all the way. It also was extremely loud because of the big clearance; the valves would clatter. So we had a series of those all happening in the first two or three days. But they fixed that in the plant right away.

Then the rocker-arm balls squeaked like mad for a while. That was because the manufacturing tolerances of the rocker ball and socket weren't within the final limits. So they made some material and coating changes and some shape changes and fixed that. These were all in the first several weeks.

And then the third thing was that we'd apparently gone too far on cylinder-bore smoothness. The rings wouldn't seat for the first 10-20 miles, and they'd burn oil like the dickens. In addition, the rings would squeak. One of the fixes, and I don't think we'd get by with it anymore, but they used to pour a little Bon Ami down the carburetor, and it would scratch the bores up.

But the plant quickly revised its cylinder machining procedures, so I'd say that, by the time a month was up, most of the teething problems were gone. After that the engine became quite successful.

L: Any problems with any other parts of the car? Or the trucks?

M: Nothing significant, really. They were both was quite good.

L: And, of course, now we remember the 1955s as models of absolute reliability and durability.

M: There was one model of the 1955 pickup line that I thought was especially interesting. Chuck Jordan, who was just a young stylist at the time, designed it, and Ed Cole pushed it through, just as he had pushed the Corvette. I'm talking about the introduction of the smooth-sided pickup truck launched by Chevrolet that year, the Cameo Carrier. [The bed] outer panels were fiberglass. The Cameo Carrier led the way to all these popular pickups we see nowadays.

Engineering the 1955 truck line was actually a bigger job than the car, because there were so many models. The truck had 75 models that year on 15 wheelbases.

L: There was obviously some effort to make the 1955 trucks look like the passenger cars; themes like the headlight brows and even the grille.

M: And the truck went to that wraparound windshield at the same time as the car. Very much alike. ❑

And here's one of the first Bel Air convertibles ever assembled (2/54). It was used for advertising and brochure photography.

Four-barrel V8, with tags showing part numbers and component release dates, was likely disassembled and used for shop-manual illustrations.

Walter R. Mackenzie

Performance engineering liaison, 1955 Chevrolet.

Walt Mackenzie was born and lived his early years in Fifeshire, Scotland. His family moved to Canada and then Pennsylvania when Walt was 14. In 1929, he joined Chevrolet at the GM Proving Grounds in Milford, Mich., taking engineering courses at night at the General Motors Institute in Flint. Walt subsequently became supervisor of a test-reporting group at the Proving Grounds and, in 1943, moved to Detroit to head up a Chevrolet wartime project.

In 1946, he was put in charge of the division's product information department which, by the mid-1950s, had become involved in performance activities as well as reporting on them. Mackenzie was made assistant director of Chevrolet public relations in 1963. He took early retirement in 1972 and now lives in California.

Interview date: 24 Jan. 1975.

Lamm: Can you shed any light on what happened when Ed Cole was offered the position of Chevrolet's manufacturing manager and ended up becoming chief engineer?

Mackenzie: You mean general manager? I believe they offered him the position of general manager.

L: No, chief engineer...before he became general manager.

Mck: Well, yes, I think I know the story. Charles E. Wilson and Harlow Curtice and L.C. Goad—Louis C. Goad—were the three top men in the corporation, in that order. And I constantly get a little thrill out of the smartness of GM executives. When they bring someone in to talk to them about an important assignment, they don't just have one of those fellow talk to him. They have all three sitting there, so that this fellow knows that whatever they offer him—and his response—it'll be accepted in a very responsible way, because here sit the top three men.

Well, that's exactly what happened on this case. Cole had made such a tremendous reputation as chief engineer at Cadillac and then as works manager of the Cleveland tank plant that they, then, faced with an era of stodginess at Chevrolet—in spite of the fact that the division was tremendously successful financially—decided it would be a nice idea to get some new blood into the organization, especially young, exciting blood like Ed Cole's.

So they brought him up to Detroit and sat him down as I've indicated. And it was my understanding...in fact, I would have bet money that Ed Cole told me this...that they offered him the position of general manager. I know Ed Cole told me that they wanted him to take over the division, which was then headed by Tom Keating, an ex-sales manager. Keating, bless his heart, knew an awful lot about marketing and dealers, but he knew very little about the technical end of the business: manufacturing, engineering, and styling.

So my recollection is that...and Cole told me personally...see, I was very

Power Pack for 1955 included twin exhausts in addition to 4-barrel carburetor. This particular car also came with overdrive.

Accessory sunvisor was a dealer option. Stainless exhaust extensions came with Power Pack but differed from those shown here.

close to him for the first three years that he was with Chevrolet...he said, No, I don't like the Chevrolet product and where it's heading. I would much prefer to get into the engineering organization and reorganize that; reorganize their product plan.

They bought that, the three of them, with the stipulation that after Cole was satisfied with what he had done, he would take over the division [as general manager].

As for Ed Kelley, he was Chevrolet's chief engineer and a man everyone held in high respect. He'd come up through the organization and had spent a great deal of time with the manufacturing people. Now there's a position in the engineering department called production engineer. This is the fellow within the design engineering department who's the liaison with the manufacturing people. The object is to marry the designs to the manufacturing process so the manufacturing becomes as painless and defect-free as possible.

As a lot of his career had been spent that way, I, among many others, thought it was just a brilliant move—although it was practically unprecedented—to put Ed Kelley over there as general manufacturing manager, because he was moving over the heads of many aspiring manufacturing people who'd been there for a long time.

L: How do the two titles stack up in rank: chief engineer versus manufacturing manager?

Mck: You can generally say they're level.

L: What about the competition end of the V8? I've been told that the 265-inch displacement was chosen, at least in part, on the basis of hydroplane class rules. Was the 1955 engine also conceived with the horsepower race and stock-car racing in mind?

Mck: It was certainly conceived as competition to Ford's flathead V8,

One reason V8 emblems didn't mount on center of decklid was so optional continental kit wouldn't hide engine identification.

because Cole had always been interested in racing. I can't say he had definite plans for Chevy to get into stock-car competition with that engine, but I can say he intended this new and very sophisticated engine to be something that was easy for the competition people to modify; something they'd really fall in love with. It was a lightweight, high-heat-rejection engine, quite simple, and as such it proved a tremendous success.

But no, I wouldn't say he was planning to get into stock-car racing when he did that engine. First and foremost he wanted to revitalize the Chevrolet product, and a very important part of that, to him, was to have a good little V8 engine. Also, by that time he'd been through the history of getting a better lightweight engine for Cadillac, and he was also very familiar with the Olds Rocket 88. He and Harry Barr had many ideas about what to do to make a new, small V8, including the stamped rockers and so forth.

L: Now what about Pontiac? How many of those ideas were shared, and what was the policy about competitors sharing innovations?

Mck: So many of those things begin at GM Central Engineering Staff and GM Research, and then they're bought by somebody. That buying or bidding has to be approved by top corporate management in view of the needs of the division.

L: And you mean the word "buying" literally, don't you? I mean, the divisions actually paid for that development work....

Mck: Well, yes, but it's all paid for initially on a sort of *pro rata* basis. All the divisions kick in. Then, when one division gets terribly interested in something...say like Ed Cole wants to get involved in a project's development within GM Engineering Staff. The goal is for his people at Chevrolet to take it over. Then, yes, they do begin to subsidize the project or the idea that way. They put a couple of Chevrolet engineers in with the

A justifiably proud Ed Cole beams down at his baby. This car catapulted Cole into GM's presidency.

Central Engineering group. Then when it's ready, the Chevy engineers move it over to Chevrolet and take it from there.

In the early stages of development, it doesn't matter who develops what, because it's all thrown into the common pot, which is GM Engineering Staff.

L: Let's get back to the competition end of it. How did that get off the ground?

Mck: Well, in those days, stock-car racing got quite a lot of attention. Fellows like Marshall Teague, driving Hudson Hornets, and Herb Thomas in Hudsons and Oldsmobiles, Fonty Flock, Tim Flock, and Jim Reed in Poughkeepsie on short tracks...those fellows were competing all the time. They were obviously very hungry for anything new that came along.

The first thing we knew in Chevrolet...we discovered that these fellows had been buying up some of our new cars and new engines and going racing with them. They were such high-revving engines, and they were running on the shorter tracks in second gear and winning.

Well, it wasn't long before people like Smokey Yunick started calling Ed Cole and saying, Hey, this is really something! Now Herb Thomas was one of the first guys we took a look at, and Marshall Teague, too, and the Flocks got into the act very early. But the car led the way. It wasn't us leading them. The racers came to us.

Now I have to put this in a peculiar kind of way. I don't want to say this is off the record about starting to subsidize racing, but yet it had to be treated discreetly, because there was a longstanding policy at General Motors that you don't go racing. And you don't help people go racing.

Well, at that time, you know, we did help some people go racing. We went from there, in early 1955, and quickly got organized. It was always discreet, never directly or openly subsidized by the corporation. Anyway, that did take place, and parts began to go out, and a few engines began to go out, because it became obvious that with just a little bit of help, these guys could really do a lot more than they were doing.

Some of the big help took the form of suspension parts, because the cars were being subjected to such tremendous forces running around these little tracks that the normal production suspension was breaking. Springs were breaking, wheels were breaking, and hubs were breaking. Which put us on the defensive, because cars that weren't prepared were breaking down and rolling over, and Chevrolet was looking like the devil. The engines were so good that the cars were being driven over their heads.

Anyway, that sort of thing kept increasing, up to the point where, in the middle of 1955, Ed Cole put me in charge of that type of thing. I was in Engineering but sort of a technical adviser to the sales, advertising, and public-relations groups, and it began to be very obvious that the effect of competition was going to be very prominent to those groups.

L: How did people like Duntov and Mauri Rose fit in?

Mck: Mauri Rose was the first man Ed Cole pulled into his office, and he told Mauri, in effect, You better get together with these guys—the Flocks and Herb Thomas and Jim Reed—and go out there and see what they really need. Then we'll fix it up so they can get it.

Well, Mauri did that, and then after he'd made some investigative field trips, he came back with a report. Cole called a meeting, and I was

Hard to believe, but traditional Chevrolet buyers hesitated several weeks before accepting the 1955's more youthful appearance.

involved. Cole said, Now this is what we want done, and Mauri's got the information, and you take charge of it; set up the budget, administer it, and so on. So that's how Mauri got into it. He worked for me in my so-called racing promotion group.

Then Duntov, who'd come into Chevrolet in 1953, into the engineering research and development group...worked for Maurice Olley [suspensions] and later for Bob Schilling. Duntov became very interested in the Corvette; wanted to devote all his time to the Corvette. He eventually got the V8 engine into it and the 4-speed transmission.

So anyway, as things went along, Duntov became more and more interested in the competition thing, and his influence broadened from just the Corvette into stock cars and so forth, because it turned out he knew a lot about performance engines and transmissions and suspension—the whole gamut of things a car has to undergo in order to run in competition.

So then, jumping over to his first, you might say, splurge into public notice, in September of 1955, the ad agency brought up the idea of sending an advance 1956 Chevrolet out to Pikes Peak to break the longstanding hillclimb record. I was put in charge of that, and I selected Duntov because I'd become friendly with him and found out what he could do. I knew he'd been a hillclimb driver, among other things, over in Europe, as well as a motorcycle racer. He also ran at Le Mans in the lighter classes in Porsches; in fact, I believe he won his class twice.

So I said to him, How would you like to run this hillclimb at Pikes Peak? Well, he jumped at it and, as a result, we mounted a tremendous promotion around that feat. So that's how Zora got into it.

Meanwhile, he'd become well acquainted with Ed Cole, and Zora had a lot of promotional ideas. That eventually led to his getting permission to go to Daytona with John Fitch and Betty Skelton to run up and down the sand and set new speed records in the Corvette. Later on we got into the development of the 1957 Sebring Corvettes, the SS and so forth. ❑

Donald H. McPherson

Senior project engineer, 1955 Chevrolet.

Don McPherson was born in Wayne, Alberta, Canada. He grew up in Detroit and graduated from Cass Tech High School in 1940. He immediately began working for Chevrolet as a drafting trainee and simultaneously took night classes in engineering at Lawrence Institute of Technology. When WW-II broke out, McPherson enlisted in the Navy and, after the war, resumed with Chevrolet, meanwhile completing his bachelor's and master's degrees in mechanical engineering at the University of Michigan.

His first postwar position was as a Chevrolet engineering design group leader, and he soon became a senior project engineer on the 1955 passenger car. Don became Chevrolet's chief engineer for passenger cars in 1966 and director of engineering in 1972. He served as president of GM Canada from 1975 until 1978, when he became general manager of Buick. McPherson retired as a GM vice president in 1985.

Interview date: 23 Jan. 1975.

Lamm: As I understand it, the novel features of the 1955 Chevy V8 were its light weight, the green-sand casting process, the stamped steel rocker arms, and the hollow pushrods. Where did those things come from?

McP: The thinwall castings came basically from GM Manufacturing Staff. They also proposed the green-sand valley core.

L: How about the stamped rocker arms and hollow pushrods?

McP: I think that was Engineering Staff.

L: Did any of that come from Pontiac?

McP: I can't help you there.

L: Were the details of that engine hard to work out?

McP: Well, I'll tell you the areas that gave the most concern. When you enter a green-sand core for the valley, you had to develop an all-new sealing technology for the cylinder heads and the intake manifold. That required quite a bit of work. The lubrication system topside also took an awful lot of development work. There must have been 30 or 40 ideas tried there to get the right amount of lubrication to the topside. Eventually they went with the wafer valve in the hydraulic lifter.

L: Was that Papenguth's idea?

McP: Papenguth and Diesel Equipment.

L: Tell me about the casting precision.

McP: The casting process was a major breakthrough. That didn't become an extremely major development, though, until we went to the 4-inch bore. Then it was *really* a thinwall casting!

L: Did you foresee such a long life for the small-block V8?

McP: Well, I don't believe any of us foresaw the future. But that particular engine set the trend for every V8 in the industry; they've all basically copied it. Take a look at the cylinder-head design. That was the lightest cylinder head for the time and probably still is. It was a very unusual design, with a minimum amount of coolant and excellent flow out the port. That's set the trend, too.

L: What do you see as the useful life of this engine? In your current position [Chevrolet chief engineer in 1975], you must certainly think about that.

McP: I think about it quite a bit. I believe the useful life of this engine will depend entirely on fuel availability. The small-block started out at 265 cubic inches, and we currently have one in production at 262 inches. It makes a very fine 262.

L: How does it do even smaller?

McP: You can pull it down as small as you want, but if you get much smaller than 262, you begin to lose the benefit of this engine. You should make the whole thing smaller. When you get down below about 250 cid, you might as well go with a Six.

L: In-line or a V6?

McP: That depends on the packaging of the automobile.

L: This engine has gone through fuel injection once already. Will it again?

McP: Nope, not at the present time. Let me tell you what the design philosophy was when this engine was put on the boards. At that time we were working on Ed Kelley's 231 V8. It was a heavy, conservative, basically square engine.

Cole's direction was that he wanted a big-bore, short-stroke engine of 265 inches with absolutely no capacity for expansion. That engine was going to be 265 and designed right to the Nth degree. But that engine did get bigger. The 283 was easy. It's been a 265, 283, 302, 307, 327, 350, and 400. The ultimate size is 400. Same basic cylinder case and heads. So you can see that the technology at the time said 265. Subsequent technology brought it up to 400.

L: But didn't they have to build that meat in from the beginning?

McP: Not by intention. It was built in by technology. The technology has been basically in the piston business, because we never thought we could build a piston with that small a compression height and still work against oil consumption.

We went to 400 inches by siamesing the bores, which wasn't considered practical in the beginning. We also got there by improving the technology of head gasketry. That technology wasn't available when we laid out the 1955 V8 but was developed subsequently for other engines. We also went from chrome rings to moly-faced piston rings. So again it was technology. The techology improved in pistons, rings, gaskets, bearings, and all those things that were developed through the years helped expand the capacity of that engine. Plus, of course, casting it with the siamesed bores. ❑

Chevrolet's advertising agency, Campbell-Ewald, began shooting for dealer literature and magazine ads in the spring of 1954.

Paul J. King

Development engineer, 1955 Chevrolet.

Born in Chicago Heights, Ill., Paul King received his BSME from the Illinois Institute of Technology in 1944 and his master's from Purdue. He worked as a designer of automatic transmissions at GM Engineering Staff from 1947 until 1953, then became a Chevrolet development engineer. King worked his way up through test, experimental, and advanced engineering to become the division's chief engineer of passenger cars in 1970. Paul became Chevy's director of engineering in 1980, then CPC director of technology centers in 1986. He retired from GM in 1987 and is currently program manager for Creative Industries Group, Inc., in Detroit.

Interview date: 8 Jan. 1991.

Lamm: Tell me how you became involved with the 1955 Chevrolet.

King: I'd been working in Central Engineering on automatic transmissions along with Frank Winchell. Then I moved to Chevrolet in 1953, and the 1955 was *the* item at that time. I started out doing engine cooling work, and it expanded from there into general development.

L: Did you have any headaches getting the 1955 to cool right? Or were there other problems that you remember?

K: There were a few things wrong with the '55 as it came out. The original V8 was designed with a cast-iron exhaust crossover in front of the engine. Very close to production we finally gave up on being able to make it live. It cracked. In other words, the heat expansion plus having it tied down on both heads made it crack. And we just couldn't lick that.

L: Why would you want to do it that way? Why was the crossover cast iron?

K: I don't really know. Chevrolet had a lot of foundry capacity; maybe that was it. Anyway, the alternatives weren't so good either. Cole had us put in a steel-tube crossover, and that had to have a slipjoint, and the slipjoint leaked. We had a helluva time making it stop leaking.

We also had some cooling problems. The engine was designed to get the hot gasses out of the combustion chambers in a big hurry. That gas would come out of the combustion chamber and was in the manifold awfully quickly. It wasn't cooled. But the heat valve...the materials we'd been using before for the heat valve had been good for 800°-900°, and in this V8 the temperature at the heat valve was running 1400°. It just burned them up in a hurry, fast as you could put them in. Now we solved the exhaust problems by having two takedowns in the rear that are steel and joining them somewhere fairly far back.

There was also clogging in the carburetor stove, in the intake manifold. There were warmup problems as the cars got older.

And in some of those early production engines, the cam bearings rotated and shut off the oil to the upper lube passages. That way the system wouldn't feed the hydraulic valve lifters.

There were some engine-mount problems. Cole had a new kind of engine-mount system.

L: The engine-mount problem persisted for quite a while, didn't it?

K: Yes, both front and rear. We got the rear ones fixed pretty quickly, because when the rear ones went, the engine sagged. Anyway, those were some of the developmental problems.

L: That's interesting, but I hadn't heard those. Another one that Alex Mair mentioned was that the rocker studs weren't pressed in tight enough sometimes.

K: Yep. That was a baddy. When you'd high-rev the engine, you'd pull the studs.

L: The other had to do with rocker-ball squeak.

K: Oh yeah. That surely was one, and that *was* annoying. But, you know, that was a very different valvetrain. I remember Smokey Yunick later, when we started to have racing successes, he told me, I don't know why that damn engine runs like it does. It's got a real [awful] valvetrain. But, of course, it did allow more freedom of valve angles.

L: Another thing Mair said that interested me was that there were very few engineering meetings, not much paperwork, and things got decided and built on the spot. Which is so different from what I see at GM today. The subject came up because I asked Mr. Mair whether he thought Mr. Cole kept an engineering log or a work diary. Mr. Mair said he didn't think so.

K: He probably didn't. Cole had a very good memory. He was really involved. One of the things that's very different is, you had a boss. Cole was the boss. He could make decisions. He didn't need big committees. And he was a remarkable guy in his capacity for work. Every Saturday

In July 1954, Chevrolet dealers flew to Detroit from all over the country to preview and learn about the new models they were about to sell.

we'd be out there, and the fact that he was constantly making decisions forced the engineers to be around him if they wanted to get their oars in.

L: There must have been rides and drives that you guys went on.

K: The first year I worked for Chevrolet I was away from home more than six months. So yeah, we were out in the field. But the rides and drives weren't big.

L: Did GM's Desert Proving Grounds near Mesa, Ariz., exist at that time?

K: Yes, the facility existed in 1953 when I came there. The buildings were very small, and there weren't individual garages for the various groups. It wasn't air conditioned, and the track was very different. It was only about half the size it is today. The 2.5-mile straightaway wasn't even a dream.

They had rains, and there was a 2-lane mud road going out from the town of Mesa to the Proving Grounds. When that mud got wet on top, the water didn't sink in, and it was like driving on glare ice. So if anyone saw rainclouds coming, they'd sound a siren at the Proving Grounds, and everybody took off for town. If they didn't start early, they wouldn't make it.

L: Any anecdotes from those drives?

K: It was all 2-lane roads. Driving was very aggressive, particularly when Ed Cole was along. Most states didn't have speed limits, so we drove pretty fast. We'd have four or five cars in a convoy passing a lot on 2-lane roads, and these were unfamiliar roads. You weren't sure how deep the gullies were over the next hill. And a lot of canyon-road driving. But there were no bad road accidents that I can remember during that period. I remember several later, but not then. Altogether I don't know how we survived.

L: Did you test the new air-conditioning system on some of those rides and drives?

K: Yes, we did.

L: Tell me about that system.

K: Well, this was the 1955 system in a 1953 Chevrolet mule. The 1955 Chevrolet marked, it seems to me, the first production usage of a front-loaded, front-mounted, totally front-contained system. Chevrolet was attracted to it mostly because of the anticipated production advantages. It was felt, because of Chevrolet's position as the corporation's price leader, that sales of air conditioning would be fairly low, probably no more than five percent at first.

Because of this, it would have been impossible to install the trunk-mounted General Motors system on line. The 1955 Chevy system was engineered, though, to be received at the assembly plant as a complete unit, totally refrigerant-charged. It was packaged so the entire unit could be dropped in on the assembly line, with the condenser slipping down in front of the radiator, compressor to be mounted to the engine, and the box with the evaporator and heater core mounted up against the cowl.

As it turned out, the forward air delivery configuration had some real advantages that we didn't fully appreciate at the time, the main one being that cold air immediately blew on the driver's and passengers' faces, so they had the sensation of being cooled instantly. This was something we later designed in, but at that time it just happened. ❑

M.S. Rosenberger

Development engineer, 1955 Chevrolet.

Rosey Rosenberger modestly describes himself as a nuts-and-bolts mechanic. He grew up in Iowa and went to a little Wesleyan school in Lincoln, Nebr., took a year of engineering there, got his first job overhauling railroad brakes in Nebraska, then assembled Fords in Omaha, and went to work as a mechanic at GM in 1927 rebuilding LaSalle V8s. He helped with the development of the 1930 Cadillac V16, worked on GM's early automatic transmission developments, including an infinitely-variable type. After the war, Rosenberger helped Cole and Barr with the Cadillac V8. Later he worked in GM Central Engineering. He joined Ed Cole at Chevrolet as an experimental engineer in 1953. His duties subsequently included Corvette and passenger-car chassis development, 4-barrel carburetion, and improving Chevrolet's big-block V8s. Rosey retired in 1965.

Interview date: 8 Jan. 1991.

Lamm: Did you work on the '55 engine right away?

Rosenberger: No, I wasn't on the engine. Cole got me over there to follow the Powerglide transmission. Chevrolet had embarked on automaticking the manual-shift Powerglide. That was designed by GM Engineering Staff. I was with Oliver Kelly's group when we did the Powerglide and the Dynaflow, and Chevrolet and Buick later took them on.

Cole asked me to come over to Chevrolet when he came in and asked me to take that job on and finish it up. Ed Kelley had started it, and we'd been working on it some from Staff Engineering, putting on a governor and doing what we had to do to make it a decent-shifting job. So that's when I went there.

You see, Paul King and I were in the same group in Oliver Kelly's gang. When I got over there [to Chevrolet] and we needed some more people, why, I got Paul over there. Paul came over ostensibly to work on the automatic transmission, because he had the background, but Cole decided he wanted a cooling engineer, and he assigned King to that job. Paul King really did that like a professional, because we had a lot of cooling work to do, especially considering all the different vehicles that engine was used in. And the 6-cylinder also.

L: Tell me about Ed Cole. What do you remember about him?

R: Well, he was a little of everything. He was good at anything. His enthusiasm and his application of thought processes to whatever was bugging him, whatever problem needed solving, he would just ride the thing right down until he either found a way to do it or he decided it wasn't worth doing. He was a tenacious man by nature, a good-natured person, and I always thought was fine to work for.

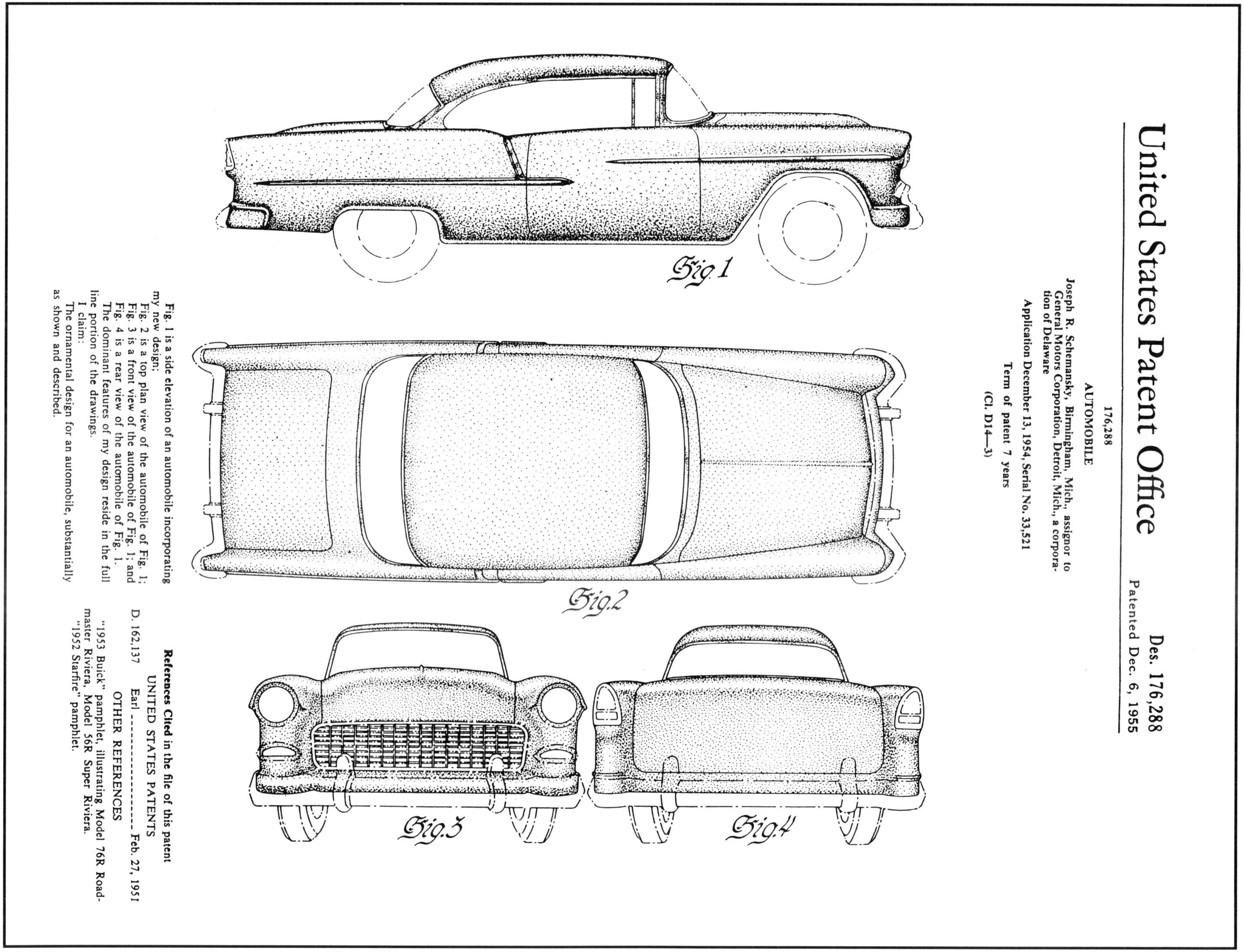

United States Patent Office

Des. 176,288
Patented Dec. 6, 1955

176,288

AUTOMOBILE

Joseph R. Schemansky, Birmingham, Mich., assignor to General Motors Corporation, Detroit, Mich., a corporation of Delaware

Application December 13, 1954, Serial No. 33,521

Term of patent 7 years

(Cl. D14—3)

Fig. 1 is a side elevation of an automobile incorporating my new design;
Fig. 2 is a top plan view of the automobile of Fig. 1;
Fig. 3 is a front view of the automobile of Fig. 1; and
Fig. 4 is a rear view of the automobile of Fig. 1.
The dominant features of my design reside in the full line portion of the drawings.
I claim:
The ornamental design for an automobile, substantially as shown and described.

References Cited in the file of this patent

UNITED STATES PATENTS

D. 162,137 Earl ---------------- Feb. 27, 1951

OTHER REFERENCES

"1953 Buick" pamphlet, illustrating Model 76R Roadmaster Riviera, Model 56R Super Riviera.
"1952 Starfire" pamphlet.

Overall body design patent was applied for at a relatively late date and was granted just after the 1955 models went off sale.

Detroit sculptor Marshall M. Fredericks designed a 300-pound, 24-foot-long, 3-dimensional 1955 Chevrolet for use on 10 billboards.

Fiberglass bas reliefs of Fredericks' sculpture appeared in Detroit, New York, Cleveland, Boston, Los Angeles, and Oakland, Calif.

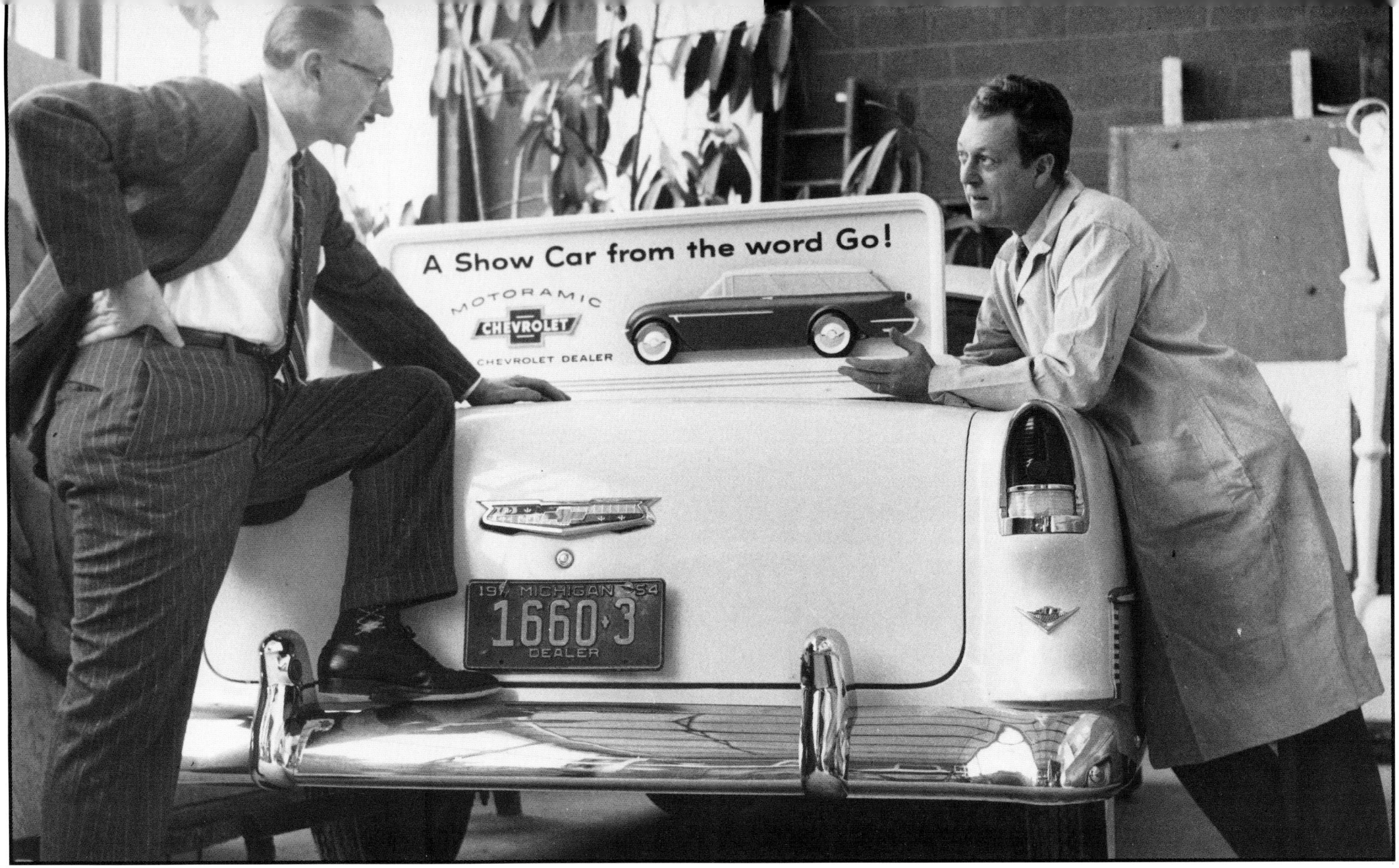

Campbell-Ewald adman Halsey Davidson (left) conceived 3-D 1955 billboard, also discussed Motorama Corvette version with Fredericks.

Some fellows thought, Well, he wants to tell you how to do everything, but I always said to them, He'll tell you if you don't dig in and start doing something on your own. He'll tell you all right, because he'll size up a problem about as quickly as anyone else coming in cold, and he was able to...no matter where he was working...he was able to stand out in comparison with other people around him. His all-around judgment, ability to do things with his hands, think and direct people, all were superb.

L: Did you go on some of the 1955 Chevrolet ride-and-drives?

R: Most of them. Cole made a point of being on them, too. Anything that was new, he wanted to go on the shakedown long enough to understand where he thought it stood in its development and whether it needed further work. So he went on darn near all the trips. Maybe he wasn't very long on a trip, but he'd be there long enough to see what it was all about. He did that at Cadillac, too. I was on some of the trips with the Cadillac gang before Cole went to Chevrolet.

L: Were these ride-and-drives mostly out of GM's Milford Proving Grounds or the Desert Proving Grounds?

R: Neither Milford nor the DPG had the facilities they have today. You used them when you arrived there, but you didn't make them your headquarters. So Ed Cole made a practice of taking engineers on his trips. In the old days, the Chevrolet gang would take one or two top engineering dogs and the mechanics and drivers, and away they'd go, whizzing around the country.

L: You mean all around the country?

R: Wherever they decided to go. ❑

Index

Appreciation

The following people and institutions helped make this book possible: Zora Arkus-Duntov, Arthur N. Armstrong, Augustine Brass Casting Co., Harry F. Barr, Duane L. Bohnstedt, Robert A. Cadaret, Pat Chappell, Chevrolet Public Relations, Prof. David E. Cole, CPC Engineering, Creative Printing, Cathy De Rosa, Letha Dolza, Edward J. Donaldson, Marshall M. Fredericks, Free Library of Philadelphia, General Motors Design Staff, General Motors Photographic, Fred Gordon, Dick Hanson, Gary Hedges, Louis G. Helverson, David R. Holls, Dave Hudgens, Harold C.L. Jackson Jr., Gene Jendrasik, Jim's Chevrolet Parts Co., Floyd Joliet, Eric Johnson, Charles M. Jordan, Joe Karshner, Paul J. King, Ralph Kramer, Joe Krachtowill, Clayton B. and John Leach, Ed Lechtzin, Lintas:Campbell-Ewald Co., Clare M. MacKichan, Alex C. Mair, Jim and Barbara Matus, Archie D. McDuffie, Malcolm R. McKellar, Walter R. Mackenzie, Donald H. McPherson, Kelly A. Morrison, Don Postma, Ellis J. Premo, Carl H. Renner, Maurice S. Rosenberger, Kari St. Antoine, Joseph R. Schemansky, Susan B. Stepek, Jim Strohschein, Typeworks, Stanley R. Wilen, Frank J. Winchell, and Pete Wozena.